AF710208

Wind Power Equipment

Wind Power Equipment

D.F. WARNE

ERA Technology Ltd
Leatherhead, UK

LONDON NEW YORK

E. & F.N. SPON

First published 1983 by
E. & F.N. Spon Ltd
11 New Fetter Lane, London EC4P 4EE
Published in the USA by
E. & F.N. Spon
733 Third Avenue, New York NY 10017

© 1983 D.F. Warne

Photoset by Enset Ltd
Midsomer Norton, Bath, Avon
Printed in Great Britain at the
University Press, Cambridge

ISBN 0 419 11410 6

All rights reserved. No part of this book may be reprinted, or reproduced or utilized in any form or by any electronic, mechanical or other means, now known or hereafter invented, including photocopying and recording, or in any information storage and retrieval system, without permission in writing from the Publisher.

British Library Cataloguing in Publication Data

Warne, D.F.
Wind power equipment.
1. Wind power—Equipment and supplies
I. Title
621.4'5'028 TK1541

ISBN 0–419–11410–6

Library of Congress Cataloging in Publication Data

Warne, D.F.

Wind power equipment.

Includes bibliographical references and index.
1. Wind power. I. Title.

TK1541.W37 1982 621.4'5 82–10689
ISBN 0–419–11410–6

Contents

Preface

In a rapidly developing field such as wind power, an isolated view very quickly becomes out of date and loses perspective. The views of co-workers in the wind power field have been freely sought in the preparation of both the text and the reference section for this book and their comments and advice are gratefully acknowledged. In particular I would like to thank colleagues at ERA, especially Mrs Elizabeth Moughton and Mr Peter Calnan for their help in compiling the reference section.

The pace of development in wind power will inevitably result in the development and availability of machines during the next few years which are not covered in the present book. Comments on omissions and additions are encouraged with a view to updating the handbook when this seems appropriate. While every effort is made to ensure the accuracy of the information given in this publication neither the Editors nor Publishers in any way accept liability for loss of any kind resulting from the use made by any person of such information.

Acknowledgements could not be complete without recording thanks to my wife Jill for her forbearance during the hours of preparation work and for typing the text. Her support and encouragement have helped greatly throughout.

D.F. WARNE

Brockham, Surrey

Summary

The handbook is intended primarily to assist in the selection of wind power equipment. The emphasis is on the detailed reference section which lists and describes commercially available equipment, and classifies this equipment into wind turbine generators, wind pumps and anemometers.

The compilation of the reference section made it evident that some preliminary text providing guidance on interpretation and use of the data would be helpful. This preliminary text outlines the basic action of wind turbines, and some of the options that are available in the design and use of the equipment. The need to avoid a major text on wind power generally resulted in cursory treatment of some subjects and for more detail the reader is directed to Golding's authoritative work *(The Generation of Electricity by Wind Power* published by E. & F.N. Spon). A basic discussion of wind characteristics is included for assistance in siting of the equipment, and it is here that some discussion of anemometers and their different modes of use appears.

Over a hundred different commercial wind turbine generators and wind pumps are detailed. In the former category a data sheet appears for each individual machine, including the power–wind speed characteristic curve and in some cases drawings and photographs. For wind pumps, the conventional approach of a pumping capacity table for each family of machines has been adopted. Coverage of anemometers is not comprehensive but a very good cross-section is presented.

It has occasionally been difficult to judge whether equipment is commercially available, particularly in the fast developing wind turbine generator area. To cover prototype, development and planned machines special sections are included which list some salient features. Large machines in the megawatt class are included in this way.

List of Symbols

a = axial interference factor
A = swept area of a wind turbine
C_D = drag coefficient
C_E = energy coefficient
C_L = lift coefficient
C_p = power coefficient
C_T = thrust coefficient
D = drag force
F_N, F_T = resultant force of lift and drag forces in the direction of motion and perpendicular, respectively
h = damping ratio
I_U = longitudinal turbulence intensity
l = characteristic blade length
L = lift force
n = number of turbine blades
P = shaft power delivered by turbine
r = wind turbine tip radius
R = Reynolds' number
S = blade area
T = turbine thrust
$\bar{u}$ = mean component of instantaneous longitudinal wind speed
u' = variant component of instantaneous longitudinal wind speed
U = instantaneous longitudinal wind speed
V = wind speed/velocity (variously used)
V_0 = wind speed at datum height
V_1 = axial wind speed well upstream of a wind turbine
V_2 = axial wind speed well downstream of a wind turbine
V_B = wind speed at the bottom of the turbine sweep
V_T = wind speed at the top of the turbine sweep
WTG = wind turbine generator
Z = height above ground
Z_0 = datum height above ground
α = angle of attack (Chapter 2); power exponent defining rate of increase of wind speed with height above ground (Chapter 5)
λ = wind turbine tip-speed ratio
μ = dynamic viscosity of air
ϕ = resolution angle for lift and drag forces
θ = angle of rotation
ρ = air density
Ω = wind turbine angular velocity

1 Introduction

Some forms of wind driven equipment have been commercialized and in use for a long time. Wind pumps and a few types of wind generator were well established even prior to the various fuel shortages and crises of the last 50 years. Since the oil supply disruption in 1973 the investigation of 'renewable' or 'alternative' sources of energy has accelerated enormously and the development of wind power devices has progressed significantly due among other things to the application of modern design techniques and the application of new materials and technology.

The result of this activity is that a much wider range of wind driven equipment has reached or is near to commercialization. With a continuing uncertainty over the price and availability of fossil fuel it is becoming quite practical and proper for those who need power supplies to consider wind driven equipment as one of the available options to meet their needs.

The procedure for selecting and installing the best and most appropriate wind generator or wind pump for a particular requirement is, however, less simple than and quite different from the well established guidelines for diesel generators and diesel pumps, for instance. The variability of the wind with time and place, the differing degrees of mechanical integrity and conversion efficiency that can be expected from the wide range of equipment designs and the general lack of precedents in some applications all contribute to the practical difficulties experienced by a prospective user.

It was against this background that the handbook was conceived. Its intention is to provide the prospective user with technical data on the range of commercial wind power equipment, and also to provide guidance on the relative merits of various designs. A working knowledge of engineering principles is assumed and the book focusses upon the wind power devices themselves not upon any ancillary devices such as electrical control and storage equipment. In the latter cases it is anticipated that those familiar with the specification and use of diesel generator systems will have adequate background to deal with selection and specification issues, since they are relatively conventional areas of technology on which specialist and general advice can be accessed from numerous sources.

The reference section is split into three principal categories concerning wind turbine generators, wind pumps and anemometers. The majority of the space is devoted to wind turbine generators (WTGs), details of which are presented in the form of data sheets for each individual machine. The predominance of these machines in the reference section reflects the diversity of available designs and the considerable and growing interest in the field from manufacturers. Although wind pumps have been sold in much greater quantity and for a longer time than WTGs, fewer variations in design exist within the current commercial ranges and the information available from manufacturers is less detailed. The presentation of data is therefore limited to a table of pumping capacities for each generic range of machines.

Anemometer characteristics are presented mainly with a view to distinguishing them as regards response, wind speed range and form of output; these three features are probably the most important to prospective wind power users in determining which instrument satisfies their particular requirement.

Components such as towers, blades, generators, etc. have been omitted. Such items are available individually and a few small companies supply components specifically for wind generator construction. The present nature of the business is such that these components are generally supplied either to manufacturers of complete equipment, or to researchers and do-it-yourself enthusiasts.

In order to assist in the interpretation and use of the three categories of the reference section several background sections are included, covering wind turbine performance theory, WTG and wind pump design and construction alternatives, and a review of wind turbine site selection methods which includes a discussion of various anemometer types. The temptation to provide a minor textbook has been avoided, and the discussion is linked to the reference section wherever possible.

With a few minor exceptions SI units have been used throughout. Recognizing the greater familiarity that many readers will have with imperial units, particularly for wind speeds and pumping capacities, a table of the principal appropriate conversion factors is included at the end of the book.

2 Wind Turbines

A wind turbine, broadly speaking, is a device which uses the natural movement of free air to produce mechanical power through continuous shaft rotation. Designs use various combinations of the different parameters available, which include blade shape, number of blades, blade pitch (fixed or variable) and transmission gear ratio, in attempts to produce higher efficiency, greater reliability or reduced cost. Variations of these parameters are incorporated in a number of unconventional or novel configurations, and the main classifications are presented later in this chapter.

The performance theory of wind turbines is complex and in some cases not yet well understood. It would be inappropriate in a handbook of this nature to enter this subject in detail. However, the performance characteristics of wind turbines vary substantially from type to type and substantial changes occur with only slight modifications to design parameters. It is important therefore that the prospective user has a basic grasp of fundamental wind turbine theory and the way that performance can be described and predicted in order that equipment can be properly specified and selected.

The following section outlines this fundamental theory and directs the interested reader to a number of classical references for the original and detailed work in specific areas.

2.1 Simple wind turbine theory

The simplest model is based on a momentum theory developed over a century ago to predict the performance of ships' propellers. The adaptation of this early theory to wind turbines was undertaken by Bilau (1925) and Betz (1927), the turbine being represented by a uniform actuator disc which creates a discontinuity of pressure in the streamtube of air flowing through it. A simple representation showing the overall control volume is given in Fig. 2.1.

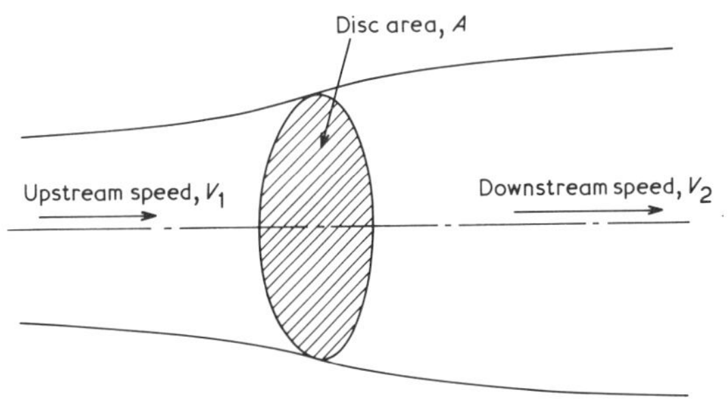

Fig. 2.1 Streamtube representation of flow through a horizontal-axis turbine.

In the simplest of models, which enables Newtonian mechanics to be applied, the flow is assumed to be steady and axial, the air is assumed to be incompressible and inviscid, and the flow downstream is assumed to be constant over this streamtube section with no pressure discontinuity across the streamtube boundary. Application of momentum and energy relationships shows that the effective average axial velocity at the disc, V, is given by

$$V=\frac{1}{2}(V_1+V_2) \quad (2.1)$$

that is, equal retardation of the flow is imposed upstream and downstream of the disc.

The power output P from the disc can be expressed as the rate of extraction of kinetic energy from this flow, that is

$$P=\frac{1}{2}\rho AV(V_1^2-V_2^2) \quad (2.2)$$

where ρAV is the mass flow rate through the disc (ρ being the air density and A the swept area of the actuator disc). It is convenient to describe the retardation in the flow using an axial interference factor a, such that

$$V=V_1(1-a) \quad (2.3)$$

and therefore

$$V_2=V_1(1-2a) \quad (2.4)$$

Substituting V and V_2 from Equations (2.3) and (2.4) into (2.2) so that power is expressed in terms of the upstream velocity V_1, the power is given by

$$P=\frac{1}{2}[4a(1-a)^2]\rho AV_1^3 \quad (2.5)$$

The terms are grouped in this way because the $4a(1-a)^2$ term is normally known as the power coefficient, C_p. This is the ratio of the actual power delivered by the disc to the free stream power flowing through a similar but uninterrupted area A. If the axial interference factor a is zero then C_p is zero and no power is developed; if a is 0.5 the downstream velocity V_2 is zero, representing the impractical case of an infinitely expanding streamtube. Maximum power extraction according to this simple theory occurs when a is 1/3 and

$$C_{P\max}=\frac{16}{27} \quad (2.6)$$

From the same set of assumptions and equations the axial thrust T on the disc can also be predicted:

$$T=\frac{1}{2}[4a(1-a)]\rho AV_1^2 \quad (2.7)$$

$4a(1-a)$ is known as the thrust coefficient C_T. C_T has an obvious maximum when a is 0.5 and the downstream velocity is zero; for maximum power output (when a is 1/3) C_T still has 8/9 of its maximum value.

The combination of the various losses which are ignored in this theory reduces the maximum power coefficient obtained in practice to claimed measured values of 0.46 to 0.47 (although it will be explained later that this is very difficult to measure accurately, and doubt can justifiably be expressed concerning the validity of some claims).

A number of workers express achieved power coefficient as a percentage of the maximum limit proposed in Equation (2.6), the term 'aerodynamic efficiency' having been used for this percentage. The value of doing this is dubious in view of the simplifying assumptions necessary to arrive at $C_{P\max}$; these assumptions raise the question of whether 16/27 really is a fundamental limit.

Attempts to refine this theory, for example by Glauert (1935) and Wilson and Lissaman (1974), include adequate representation of angular motion induced in the flow by the turbine action. This requires a departure from the simple axial flow previously described, and generally predicts a lower power coefficient due to the higher power content in the wake, which is essentially lost.

It will be noticed from Equation (2.5) that the turbine power is proportional to the cube of wind speed. Since natural winds vary quite substantially even within quite small areas and localities, the power and energy delivered by a wind turbine is (unlike many other forms of generation equipment) therefore critically dependent on equipment location. Careful siting is therefore at least as important as the design or selection of equipment with high C_p. The siting issue is discussed separately in Chapter 5.

2.2 Principal classes of wind turbine

The simple theory which has been outlined is the result of a large number of approximations and is broadly applicable to the limiting case of a turbine with a large number of blades operating at very high rotational speed with its axis aligned to the wind direction. This is unrealistic (more so for some turbine types than others) and in order to improve performance prediction through a more detailed account of real conditions immediately around the turbine it is necessary to use a different approach. Not the least of the complications in addressing this subject is the variety of wind turbine types that have been conceived and developed over the years, each needing a special and unique emphasis in the theoretical treatment to give good performance prediction. Before going on to outline some of these theoretical approaches in the following section it is therefore worthwhile describing the main types or classes of wind turbines that are of interest. Probably the most important generic distinction is between horizontal-axis and vertical-axis turbines. The former group (sometimes more accurately called wind-axis turbines) include the traditional corn grinding windmill, the multi-blade wind pump and the modern propeller-like turbine. The latter group (conversely called crosswind-axis turbines) probably have earlier origins and include a wide variety of configurations from ancient Chinese and Afghan machines through to high-speed twentieth-century designs using aerofoil sections. Functionally, in both horizontal-axis and vertical-axis groups, an important subdivision can be made between types having high and low solidity. Solidity can be defined in different ways for various turbine configurations, but broadly it represents the total blade area divided by the swept frontal area of the turbine.

In describing the principal classes of wind turbine it is

helpful to refer not only to solidity but also to the tip speed ratio (λ) of the turbine, which is the ratio of turbine peripheral speed to free stream wind speed, V_1:

$$\lambda = \frac{\Omega r}{V_1} \tag{2.8}$$

where Ω is the turbine's angular velocity and r its tip radius. Taking horizontal-axis turbines first, the high-solidity type is characterized by the traditional wind pump, having a large number of broad metal blades. This type of turbine does operate fundamentally through the principle of aerodynamic lift which is outlined in the following section, but the turbine is relatively heavy in construction, and torque or power control is difficult. These features possibly impose a relatively low upper limit on the size of turbine that can be successfully built in this class. However, inter-blade aerodynamic interference is high and imposes some inherent limitation on the torque and speed developed in high winds. These turbines operate at relatively low values of tip speed ratio, λ generally being less than 2.

Higher values of power coefficient can be achieved by making the blades of true aerofoil section (thereby increasing the lift and reducing drag) and using relatively few blades so that reasonably steady wind conditions are re-established as incident to each blade following the passage of and disturbance created by the previous one. The resulting turbine closely resembles a propeller and is often known as the 'propeller' type. Higher tip speed ratio and higher power coefficient are achieved in this way. The contrast in λ and C_p between this propeller type and the previous type with many blades and high solidity is shown in Fig. 2.2.

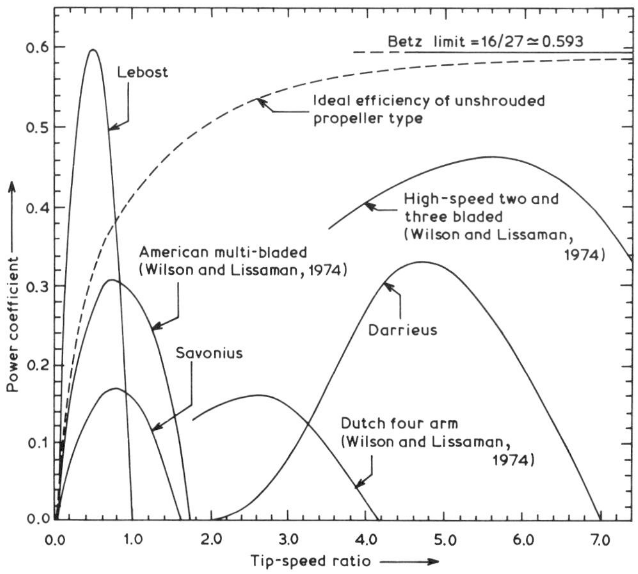

Fig. 2.2 Power coefficient versus tip-speed ratio for various vertical- and horizontal-axis turbines.

Vertical-axis turbines or panemones may likewise be classified into high-solidity and low-solidity types. The former may take a wide variety of forms, the physical action being dominated by drag effects and the blades usually travelling more slowly than the air passing them, resulting in λ less than unity.

Going back over literally thousands of years (it is almost certain that the first wind turbines were panemones), much design effort and ingenuity has gone towards means of reducing the drag created by blades returning into the wind and relying on the resulting differential drag to produce the torque. The familiar three-cup anemometer works on this principle and a few other variants are shown in Fig. 2.3, including the famous turbine conceived and initially developed by Savonius (1931).

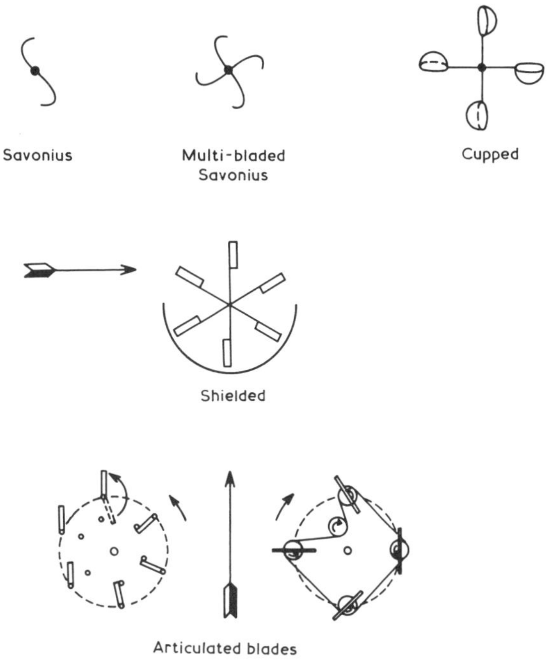

Fig. 2.3 Various types of panemone.

The arrival of the high-speed low-solidity vertical-axis turbine (like its horizontal-axis propeller-type counterpart) came with the advances in aerodynamic theory following the First World War. The two principal types are shown in Fig. 2.4. The curved blade or 'troposkien' type has the advantage of simplified and largely tensile blade stresses; the 'squirrel-cage' type has the advantage of cheap blade production, but can result in high blade bending moments. The original conceptual work on this class of machines was done by Darrieus (1931), but development of its theory and understanding scarcely progressed (with the exception of some French work in the early 1950s) until the late 1960s, possibly because of the lack of relevance to propeller work. In the last ten years considerable interest has arisen and a number of design variants have emerged. Notable is the variable pitch version of the squirrel-cage turbine (known by some as the 'giromill' or 'cycloturbine'), which uses a central wind vane with a cam and push rods to each blade, the blade being capable of rotation about its principal axis. In this way a cyclic pitch change is achieved, and self-starting can also be arranged. A further variant, conceived and evolved by Musgrove (1978), involves hinging the blade at a position away from the mid point so that, as rotational speed increases, centrifugal forces draw outwards one end of each blade.

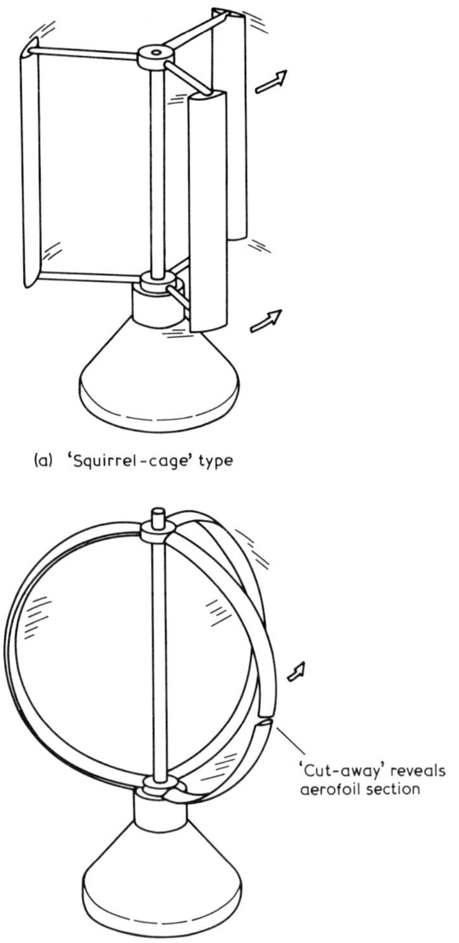

Fig. 2.4 Squirrel-cage and troposkien vertical-axis turbines.

This takes it (and the inward end of the blade) away from optimum tip speed ratio, reducing the torque and thereby limiting the rotational speed.

2.3 Lift and drag

To undertake detailed design calculations or make accurate performance predictions it is necessary to adopt a different approach to the simple theory outlined in Section 2.1. The air flow around the individual blades must be considered and the performance of aerofoil sections (well documented through propeller and wing research) must be taken into account.

The resultant force on an aerofoil section is classically separated into lift and drag components. The drag force acts in the same direction as the relative velocity of the air to the aerofoil, and the lift force is perpendicular to this. The velocity and force vectors are shown simply in Fig. 2.5.

The lift and drag forces L and D are proportional to air density (ρ), blade area (S) and the square of air velocity relative to the aerofoil (V), the proportionality constants being known respectively as the lift coefficient C_L and the drag coefficient C_D. Hence

$$L = \frac{1}{2}\rho V^2 S C_L \tag{2.9}$$

and

$$D = \frac{1}{2}\rho V^2 S C_D \tag{2.10}$$

As shown in Fig. 2.5 the lift and drag forces can be considered (except in stalled conditions) to act at a fixed point with respect to a specific blade section, and they can be resolved in the direction of motion and perpendicular to this so that

$$F_N = L\cos\phi - D\sin\phi \tag{2.11}$$

and

$$F_T = L\sin\phi - D\cos\phi \tag{2.12}$$

where F_T contributes to the torque developed by the turbine and F_N contributes to the thrust on the turbine. Although referred to as proportionality constants, C_L and C_D are in themselves complex functions. These coefficients have been measured in wind tunnel tests for a wide range of aerofoil sections, the most extensive classification being the NACA series documented by Abbott, Von Doenhoff and Stivers (1945). C_L and C_D depend primarily on the angle of attack, α (shown in Fig. 2.5). For the types of aerofoil sections typically used for wind turbines, C_L reaches a peak at about 10–15° (the angle of stall), after which it decreases. The drag coefficient is less easy to predict, being more dependent on surface roughness and friction effects. Both coefficients

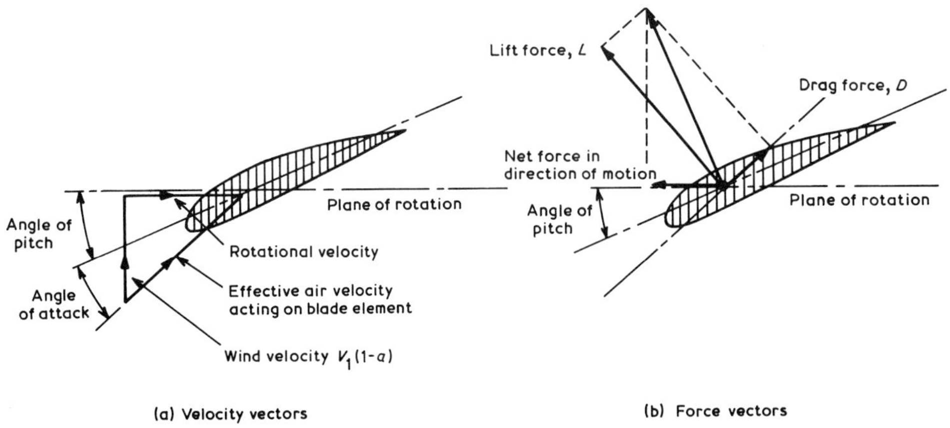

Fig. 2.5 Lift and drag on an aerofoil section.

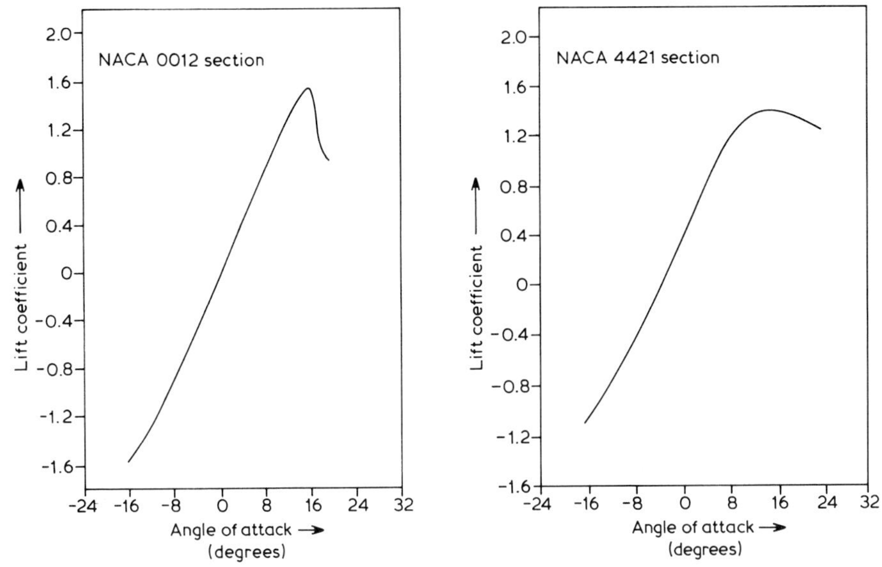

Fig. 2.6 Variation of lift coefficient with angle of attack for NACA 0012 and 4421.

are affected by the Reynolds number R (a dimensionless number associated with the flow), which is given by

$$R = \rho \frac{Vl}{\mu} \tag{2.13}$$

where μ is the air dynamic viscosity and l is a characteristic length, usually the blade chord in this case. Data for most aerofoil sections exist only for narrow bands of Reynolds number, say in the region 10^6 to 10^7, which applies to wing and propeller applications. Some wind turbines operate with Reynolds numbers almost an order of magnitude lower and although certain extrapolations can be made these data (particularly for small machines where the Reynolds numbers are lowest) must be treated with caution. Typically, C_L is plotted against angle of attack and C_D against C_L; examples of the characteristics for NACA 4421 and NACA 0012 sections are shown in Figs 2.6 and 2.7.

Using these data the torque and thrust developed by a section can be calculated. However, each section of a turbine blade is operating at a different tangential speed, and the angle of attack (between the blade chord line and the air flow relative to it) therefore changes continuously along the blade. To produce an angle of attack corresponding to maximum lift to drag ratio at all points on the blade, the designer may introduce twist into the blade. The blade may also be tapered to maximize torque production, and larger blades may incorporate different sections at different radii, reflecting the changing structural and aerodynamic functions of the aerofoil at various radial positions. A complete analysis therefore has to be built up by using a strip theory, which sums contributions to torque and thrust from a number of separate radial stations along the blade, using the described derivations of lift and drag forces.

In the high-speed vertical-axis turbine the same principles of lift and drag apply, but the fundamental action is less intuitive in nature and requires some explanation. The important distinction is that, for a horizontal-axis turbine, each blade section experiences a constant angle of attack during one revolution (assuming steady conditions), whereas in the high-speed vertical-axis case the angle of attack experienced by each blade section varies continuously through one revolution. Fig. 2.8 shows a plan section of the turbine, the blade being mounted with its chord approximately tangential to utilize best the lift forces. The velocity triangle is also shown in Fig. 2.8a to indicate the angle of attack; this triangle is rearranged for convenience in Fig. 2.8b to show the

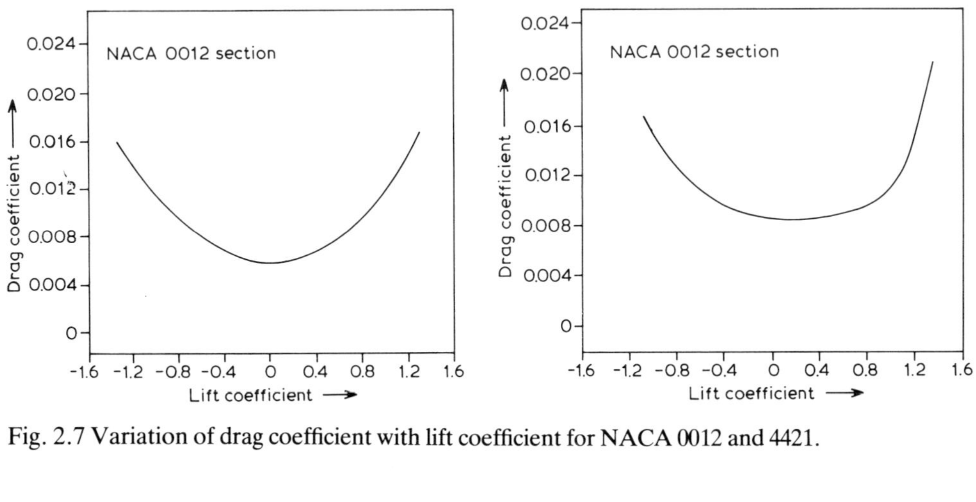

Fig. 2.7 Variation of drag coefficient with lift coefficient for NACA 0012 and 4421.

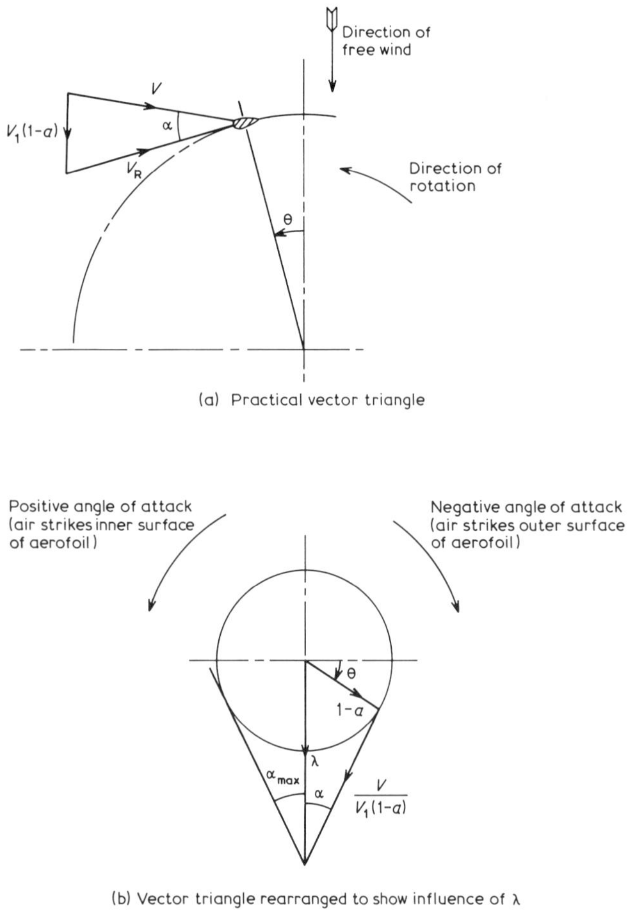

Fig. 2.8 Airflow vectors at the blade of a high speed vertical axis wind turbine.

variation of angle of attack throughout one revolution, and it is seen that

$$\alpha_{\max} = \sin^{-1} \frac{1-a}{\lambda} \tag{2.14}$$

For an angle of stall of 12° and an axial retardation factor a of 1/3 (for maximum C_p), a useful lift can be produced at all points of the trajectory (except where α is zero) for $\lambda > 3.2$. This means that in steady flow conditions the turbine is not self-starting and has to be driven by other means up to the speeds at which it produces a net torque in the direction of rotation.

A performance estimate using lift and drag characteristics can be arrived at by summation in a manner akin to that described for horizontal-axis turbines. The important and necessary refinement to performance prediction in this class of turbine is a better understanding of the wake shed by the upstream blades and the effect this has on torque and thrust developed by those blades passing through the downstream sector. This subject is presently receiving considerable attention by workers in the field.

2.4 Performance and application matching

The shaft power delivered by a wind turbine can, as in most mechanical drives, be influenced by the matching of torque–speed characteristics for the prime mover and the load. This is partly under the control of the designer, with respect to the transmission ratio selected, and (if used) the way in which blade pitch is varied, and partly under the user's influence, with respect, for instance, to the control and excitation system used on a generator or the pump size used in a wind pump system. It is therefore important to appreciate the factors which govern correct matching of the wind turbine and the connected load.

The most general description of a turbine characteristic is the power coefficient/tip speed ratio curve, such as indicated in Fig. 2.2.

From Equation (2.5) we have

$$\text{Torque} = \frac{1}{\Omega} \times \text{Power} = \frac{1}{\Omega} \times \frac{1}{2} C_p \rho A V^3$$

i.e.

$$\text{Torque} = \frac{\pi r^3 \rho}{2} \times \frac{1}{\lambda} \times C_p \times V^2 \tag{2.15}$$

(substituting from Equation (2.8)).

A family of torque versus rotational speed curves can then be drawn, for a range of wind speed values V. Using the high-speed turbine case of Fig. 2.2 and assuming unity radius, such a family of curves representing a fixed pitch turbine is shown in Fig. 2.9. Superimposed on the family of curves is a set of constant power lines in order to indicate the region of maximum power delivery for each wind speed.

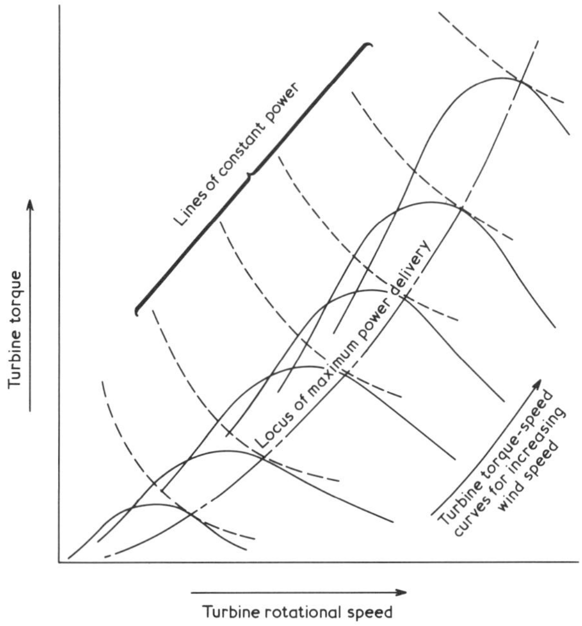

Fig. 2.9 A typical family of torque–speed curves for a fixed pitch turbine.

The load acting at the turbine shaft is represented by a load line. In the case of a mechanical load such as a pump, the position of this line is strongly influenced by the transmission ratio between wind turbine and load; the effect on the load line of 50 per cent increase and 50 per cent decrease in transmission ratio is shown. Maximum power delivery is clearly achieved by designing the transmission ratio such that the load line and the locus of points representing maximum turbine power output coincide as closely as possible. The load line can be further influenced by the user's application, such as the head in a water pumping application, or the voltage regulation and excitation control system in an electrical system. The curves shown are derived for a simple fixed pitch variable speed turbine but pitch variation is often used, particularly in small wind turbine generators up to a few metres in diameter. Pitch change mechanisms on machines in this class are often centrifugally governed using fly-weights near the hub; the effect of pitch change can be to flatten the torque–speed curves for the turbine at individual wind speeds and this can make the turbine–load matching less sensitive to selection of gear ratio and other factors.

It is clearly in the user's interest to consider optimum matching, for by this means the energy production from a particular turbine can be maximized. Power coefficient/tip speed ratio curves, gear ratios and load torque–speed characteristics should therefore be sought from the manufacturers and the application data wherever possible. It has to be noted, however, that only seldom do the turbine manufacturers produce such information and the matching calculations cannot always be made.

The matching question is different in the application where a wind turbine generator is directly connected to the electricity supply network and operates at nominally constant speed, governed by the network frequency. Here it has to be accepted that the turbine (particularly if fixed pitch), will be operating off optimum for a substantial part of the time, and correct gear ratio etc. is important for maximum energy production for the mean wind speed at the intended site. This is taken further under the discussion on energy calculations in the next section.

References

Abbott, I.H., Von Doenhoff, A.E. and Stivers, L.S. (1945), *Summary of Air Foil Data,* NACA Report 824.

Betz, A. (1927), *Naturwissenschaften,* **XV,** 905–914.

Bilau, K. (1925), *Elektrotech. Z.,* **46,** 1405–1407.

Darrieus, G.J.M. (1931), *Turbine Having its Rotating Shaft Transverse to the Flow of the Current,* US Patent 1835018.

Glauert, H. (1935), Airplane propellers, in *Aerodynamic Theory,* (Ed. W.F. Durand), Vol. 4, Springer, New York.

Musgrove, P.J. and Mays, I.D. (1978), Development of the variable geometry vertical axis windmill, *Proc. 2nd International Symposium on Wind Energy Systems,* BHRA Fluid Engineering, Cranfield, Bedford, UK, Volume 1, pp. E4–39 to E4–60.

Savonius, S.J. (1931), S-rotor and its applications, *Mech. Eng.,* **53,** 333–338.

Wilson, R.E. and Lissaman, P.B.S. (1974), *Applied Aerodynamics of Wind Power Machines,* PB 238595, Rep. No. NSF–RA–N–74–113, NTIS, Springfield, Virginia.

3 Wind Turbine Generators

3.1 The principal applications

Wind turbines deliver their power through a rotating shaft and in this respect are similar to other prime movers such as diesel engines and steam turbines. In principle then, a generator can be coupled to this shaft and the electrical power delivered can be used to serve the multitude of different purposes for which electricity is required today. In practice an important difference between the wind turbine and engines driven by fossil fuels is that the power delivered by the former is to some extent uncontrolled and unpredictable over very short periods of time. In most applications for electricity, power is normally required 'on demand' and not whenever available, so it is important to have some means of storage or reserve supply. This requirement has been one of the main limitations on the use of wind turbine generators over the years, and the applications which have been most widespread have often been those in which the cost and complexity of providing this reserve capacity or storage capability have been minimized.

The four leading applications are:

(a) Autonomous electric power supplies, where the wind generator will operate in conjunction with electrolytic batteries, and perhaps also a diesel generator and inverter;

(b) Parallel operation with a much larger power network, where the system frequency and/or pitch variation is used to control the turbine speed at nominally constant level. The network is in this case essentially the reserve capacity for periods of low or zero wind speed;

(c) Operation with a larger power network through an inverter. In this case the WTG may run at variable speed, delivering variable frequency, and a regulator and inverter are used to condition the power for delivery to the system. The network again is available to the user as reserve capacity in low wind conditions;

(d) Water heating using resistive immersion elements. In this case the WTG may run at variable speed, delivering variable voltage and frequency. Energy storage in low wind periods is provided by the thermal capacity of the hot water, and reserve capacity in the form of other water heating facilities may also be available.

Wind turbine generators suitable for each of these four applications are included in the reference section. They are being promoted for use in all of these applications with greater or lesser success depending for instance upon local mean wind speeds, alternative costs of delivered energy and the co-operation of electricity supply authorities in individual countries.

3.2 Design alternatives

The previous chapter outlined some of the main options in wind turbine design and configurations but there are further considerations which can affect the reliability and cost of the equipment and its suitability for certain environments and applications. It should be appreciated that wind turbine generators, although they have been used for many years, are still relatively poorly understood in the detail of operational conditions and stresses that they are exposed to in a natural environment. Much of the commentary on different types which follows is therefore subject to the outcome of many more years design, manufacturing and operational experience. Design features adopted by manufacturers tend to be similar within each of the four applications and the features are discussed with particular reference to the applications.

A fundamental design choice highlighted in the previous section is between horizontal- and vertical-axis turbines. The relative advantages have been outlined and it is evident that although the high-speed vertical-axis machine offers promise, its performance is not as well understood as that of the horizontal-axis machine. This early development stage is reflected in the relatively few vertical-axis machines that appear in the reference section describing commercially available machines. Most of the commentary which follows therefore relates to horizontal-axis machines, but before starting upon this the specific options in vertical-axis turbines are outlined.

3.2.1 *Vertical-axis machines*

Of the high-solidity variants discussed in Chapter 2, only the Savonius and related types are currently offered as serious commercial machines. They are represented in the reference list by one of the smallest machines appearing, the W250 produced by Elektro GmbH of Switzerland. (In fact this is not a true Savonius-style machine since the blades do not overlap in the centre.) A commonly stated disadvantage of the Savonius type of turbine is the so-called Magnus effect, which is the sideways thrust developed by the turbine because it creates an asymmetric wind flow around itself. This lateral thrust fluctuates substantially for a two-bladed machine and in larger turbines it becomes difficult to limit deflections and stresses to safe limits. Units with an outer frame rigidly supporting an upper bearing have been produced, but this frame tends to negate the advantages of the vertical-axis machine in respect of restricting the major structure to a limited area below the turbine. The fluctuating effect can be reduced by increasing the number of blades or by twisting the blades around the centre turbine axis in the form of a helix, as in a few machines being developed by small companies in UK and Canada. Numerous modifications to the basic design have been attempted; the vane and deflector system on the Elektro W250 (intended to accentuate the flow on the downwind running side) is an example. The power coefficient (C_p) of Savonius-type turbines tends to be lower than high-speed vertical- or horizontal-axis turbines, and for large devices where additional material is required to maintain stiffness and combat the Magnus effect the Savonius turbine becomes uneconomic and usually impractical.

For turbines up to a metre in diameter (rated up to a few hundred watts), where the Magnus effect is relatively insignificant, the Savonius turbine can therefore be considered as a reasonable option for practical and reliable use.

Of the high-speed (low-solidity) vertical-axis turbine variants described in Chapter 2, three principal types are either commercially available or under advanced development.

The 'giromill' or 'squirrel-cage' type is represented in the reference section by the Pinson Cycloturbine (a machine of 4.58 m diameter rated at 2000 W), and it is also at the development or prototype stage in Denmark (Dana Vindkraft) and the USA (McDonnell–Douglas). The straight blades allow cheap production, the easy change of blade length facilitating a 'family' of machines (increased output can also be achieved by mounting the same blades at a larger radius) so that in principle a range of machine sizes and ratings can be offered with minimum blade production cost. Bending stresses in the blades tend to be a significant design limit and means such as the double radial arm support in the Pinson Cycloturbine are sometimes necessary to restrict stresses to safe levels at high rotational speeds. The relatively simple structure does permit blade pitch change and this can readily be effected (as shown in Fig. 3.1) using a central cam directed by a wind vane. The cam is connected to each blade with a push rod which changes the pitch setting as the turbine rotates; this principle of operation has been outlined by Grylls (1978) and Brulle (1977). The cyclic pitch change achieved in this way results in a higher operating power coefficient since the angle of attack of each blade can be maintained nearer the optimum for a larger part of the blade trajectory. Whether the additional power and energy produced can be justified against the extra cost of the pitch change mechanism is uncertain at this stage. More importantly, the pitch change facility assists self-starting, which is a common problem in low-solidity vertical-axis turbines. It has been suggested that the giromill principle is best suited to larger machines, since in small turbines with relatively high rotational speeds the number of stress cycles in linkages and bearings accumulates very rapidly as the turbine operates. This could lead to early fatigue damage and possibly failure. It is also likely that short-term fluctuations in wind direction will contribute to the number of stress cycles, and the dynamic interaction of the wind vane and turbine is not yet fully understood. It is possible that in small machines where these effects are most prominent the best solution is to use the pitch change mechanism for starting purposes only and render it inoperative during normal running.

Commercial development of the variable geometry vertical-axis machine conceived by Musgrove (1978) is presently confined to the 6 m machine produced by PI Specialist Engineers, this being included in the reference section. In its simplest form, the variable geometry

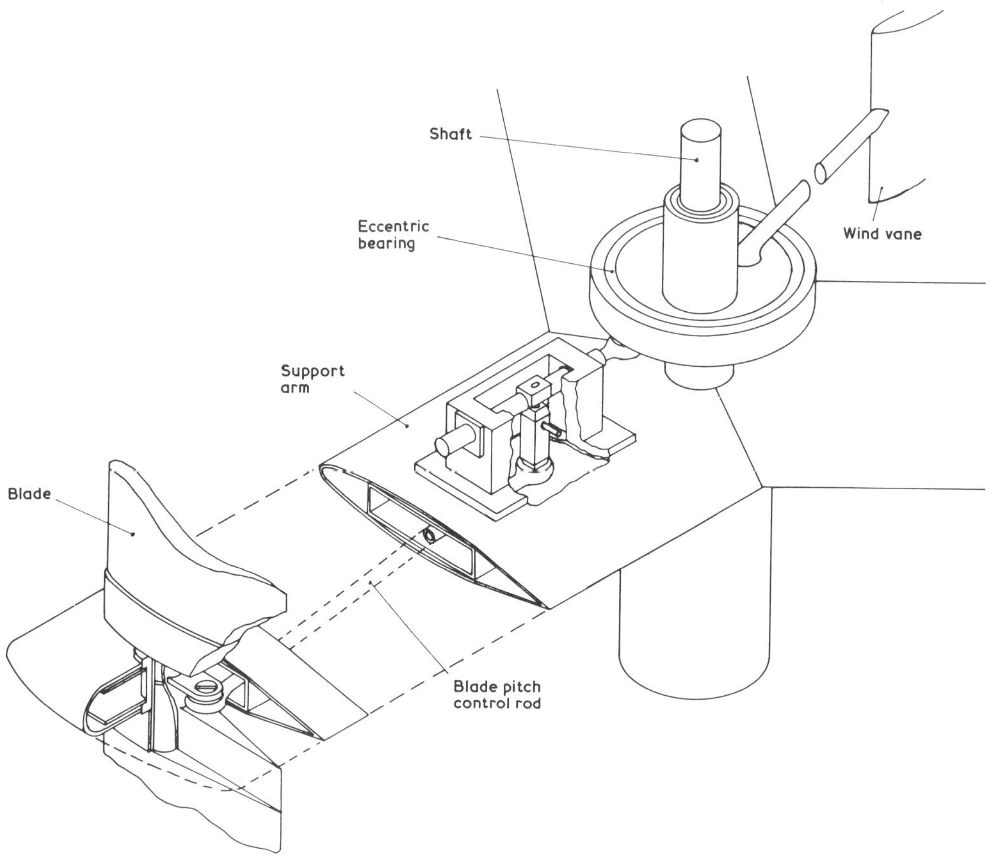

Fig. 3.1 Pitch control system for a giromill (courtesy of Exeter University).

turbine has blades which are mounted away from their mass centre, so that under increasing centrifugal forces as the speed rises one tip of each blade tends to move radially outwards. This radial movement is restrained by spring action through a connection to the outward moving section of the blade. A direct consequence of the change in blade angle at constant rotational speed is that the peripheral speed of the outer section increases, while that of the inner section decreases. In both sections the tip speed ratio and the angle of attack change all along the blade away from the previous optimum, and the torque decreases, therefore tending to decelerate the turbine and achieving a self-limiting effect. In more sophisticated versions, the attitude of the blade can be controlled independently of speed by changing the restraining tension in the spring action. This system is better suited to constant speed machines operating with the generator connected directly to the fixed frequency network; the attitude changing mechanism can then be used for torque regulation and for overspeed protection in case of loss of load or some other fault. The former system with automatic speed-dependent attitude control is suitable for autonomous battery charging or electric heating duties, and it is this mode of operation which has been adopted in the 6 m PI Specialist Engineers' machine. Like the giromill this machine has the advantages of cheap blades which are readily adaptable to different turbine sizes and ratings. The latest evolution of the variable geometry concept (Clare, 1981) introduces a principle in which the centre point of the blade is hinged. Each half of the blade has a strut and drag link which enable the blade to be positioned at a range of angles, forming an 'arrow-head' at each end of the cross arm. This design has been developed for ultimate use in megawatt scale wind turbines but in principle it could find application in small machines, depending upon costs.

A third variant of the vertical-axis turbine, which is at present the most commonly investigated and developed is the 'troposkien' type, so called because the blades have an elevation profile similar to that adopted by a spinning length of rope. Much of the recent development work on this type of machine has been pioneered by National Research Council in Ottawa (Templin, 1974) and more recently by Sandia Laboratories, Albuquerque, New Mexico (Blackwell, 1977). The classical advantage claimed for the troposkien turbine is that because of the shape the blade stresses are relatively simple and mainly tensile. In practice a number of complications such as radial blade movement, blade lag and general profile distortion arise partly because in the downwind part of the trajectory the blade is operating in the wake of the upwind blade passage, and also because of the natural turbulence of the wind. These complications have most serious consequences in larger wind turbines, where it has been found necessary to introduce cross-bracing of one form or another between the blades. The blades produced so far have commonly been of extruded aluminium section, and this accounts for the specific commercial interest of DAF-Indal, a subsidiary of RTZ (whose 11.2 m diameter 56 kW rated machine appears in the reference list) and ALCOA, who have initiated a range of machines. In earlier machines some difficulty was experienced with 'rippling' of the blade section on the inner face of the blade near the mid point, following the bending operation. This is the area of maximum contribution to aerodynamic lift and to torque and the 'rippling' is obviously undesirable because of the drag created. One way of avoiding the problem is to fabricate the blades from short straight sections, an example of this in the reference list being the Bristol Aerospace LICUS A machine. An alternative solution is to use composite materials in place of extruded aluminium; this approach was adopted by Kaman Aerospace in a contract to deliver blades to Sandia Laboratories. In order to ensure self-starting, some troposkien designs incorporate small Savonius-style turbines, occasionally called 'bucket starters'; these are shown in the LICUS A, B and C machines of Bristol Aerospace. Concern about the drag imposed by these devices at full rotational speed has resulted in the design by Sandia Laboratories (Blackwell *et al.* 1977) of a hinged unit, the Savonius turbine 'S' profile collapsing into a closed circle at high speeds under the influence of drag. More recent research on the various styles of vertical-axis turbine has shown that self-starting may be achieved by use of higher solidity, and it seems questionable whether the expense and complexity of the Savonius-type starters is cost effective or even necessary. An alternative technique is to start the turbine using the generator as a motor, this being energized following incidence of a pre-set wind speed read from an auxiliary anemometer, or using other control techniques. The troposkien turbine is essentially fixed pitch and consequently requires overspeed protection, whether it operates within an autonomous (battery charging, heating or pumping) system or a network connected system. The most common means of overspeed prevention are centrifugally operated spring-loaded flaps near the outer periphery of the blades. These are simple and effective, but may give problems of equalization on two or three individual blades and may suffer premature icing due to local effects (while the main turbine still remains ice-free and capable of producing normal torque). The well-documented failure of the 24.4 m troposkien machine on the Magdalen Islands was ultimately caused by failure of such spoilers, following a protracted period of running in overspeed.

Other serious design considerations for vertical-axis turbines, such as number of blades, fixed or variable speed, mechanical transmission and generator type are similarly applicable to horizontal-axis machines and are discussed in the following section.

3.2.2 *Horizontal-axis machines*

The main options in machine design and construction are as follows:

(1) Number of blades (commonly two or three)
(2) Blade material, construction method and profile
(3) Fixed or variable pitch blades
(4) Operation upwind or downwind of the tower
(5) Fixed or variable turbine speed
(6) Orientation by tail vane, fan-tails, self-centring action or servo-control
(7) Synchronous, induction, d.c. or other types of generator
(8) Gear, belt, chain or hydraulic transmission or direct generator drive.

These are discussed through the course of this section, with a number of further considerations connected with the application of the machines.

Most of the more obvious and probably fundamental choices relate to the turbine itself, and one of the controversial topics is the number of blades. Traditionally, small wind turbine generators have most commonly used three blades, but some have incorporated two or occasionally four or six, and more recently one-bladed machines have been the subject of serious R & D. The arguments centre on a number of technical and economic issues. For a given diameter, as the number of blades is increased (with the same overall solidity) the peak power coefficient increases, but not substantially. Typically, the increase in power coefficient when changing from one to two blades is about 10 per cent and from two to three blades about 5 per cent; this increase becomes progressively less in going from three to four blades and to greater numbers. The extra number of blades does, however, bring additional cost, and the hub (particularly for three blades compared with two) is more complicated and expensive the higher the number of blades. The extent to which the additional output and additional cost outweigh each other is not clear at this stage. A factor which favours a higher number of blades is reduction in cyclic loading. Each blade during its revolution experiences a changing wind speed, partly due to the vertical gradient (a phenomenon described in Chapter 5) and partly due to tower blockage effects. (The latter are especially important in the case of turbines mounted downwind of the tower.) This produces a once per revolution cyclic load on the blades and an n per revolution cyclic load on the hub, transmission, nacelle and tower (where n is the number of blades). Loads also appear at various harmonics of these frequencies. More importantly, the amplitude of the cyclic loads on the hub etc. decreases as the number of blades rises, since the total torque and thrust reduction due to one blade passing into a low-wind area diminishes. A further feature which favours three blades instead of two, especially in free-yaw machines, is the reduced cyclic precessional or gyroscopic torque as the turbine changes direction to follow the wind. The two-bladed turbine exhibits a dramatic change in inertia in the precessional mode as its rotational position varies from horizontal to vertical. As the orientation system applies torque to change the turbine direction, this changing inertia results in a jerky motion which is much less pronounced in three-bladed turbines. Finally, it is worth noting that for fixed pitch turbines, three blades provide considerable scope for inter-blade stiffening. Use of this has been fully capitalized upon in a number of past and present Danish machines and is evidenced in the reference list in the Kongstedmollen and Wind-Matic machines. It is evident that without a substantial body of operating experience with the different types and without reliable indicators of cost (which only come from mature production of a number of these different types), no clear judgement can emerge at this stage.

A range of choices is available in blade design and construction, the important features being profile and shape, materials and method of construction. The requirements for profile and shape have been discussed in Section 2.3. Aerofoil sections are usually employed to achieve a high lift/drag ratio and consequently high power coefficient, but there is a special requirement for a high thickness/chord ratio especially near the blade root in order to reduce the mean operating stresses and to withstand the arduous fatigue conditions in the wind turbine case. NACA 0012, 0015 and 44XX profiles have been commonly selected for these reasons. Candidate materials range in their apparent sophistication from sitka spruce to reinforced plastics. The principal alternatives presently used in small WTGs are aluminium (represented for example in the reference section by the larger Aerowatt machines and the vertical-axis machines of DAF, Dynergy, Pinson and PI) and wood (notable being the Winco, the horizontal-axis Elektro machines, Sencenbaugh, Dakota and Aeropower). The most commonly used wood is sitka spruce, although the small Aerowatt units have beechwood blades. Wood blades are often strengthened in some way using aluminium, steel or even copper. Aluminium has advantages but its properties under very long-term load cycling are at present not well understood. (10^9 cycles or more can be expected in a small wind turbine over a 20-year life. It is worth commenting here that in only a few engineering applications are components and materials subjected to such a high number of load cycles as the wind turbine operating in a natural wind environment.)

A few manufacturers are attempting to capitalize on the high strength/weight ratio of various glass-fibre composites and the promise of cheap mass production techniques using these materials; whether this economy can be achieved remains to be fully demonstrated. The non-metallic composites developed for high strength–weight ratio usually have good resistance to corrosion but their long-term fatigue properties are often uncertain and close attention to bonding is usually necessary, within the blade structure itself and to the more conventional materials of the hub.

Finally there are the various steels. Mild steel is probably best known with respect to its fatigue properties but it has a poor strength/weight ratio and requires special protection if acceptable long-term corrosion resistance is to be achieved. Stainless steels have been used but these are of course relatively expensive and tend to be used where cost is not of primary importance, for example in the Bristol Aerospace LICUS range intended for high-reliability remote operation.

The method of construction tends quite naturally to be closely coupled to the materials used. Wooden blades are either cut from solid material or built up from laminations, and metal strengthening is often added, particularly on the leading edge to reduce erosion (the Winco machine is one of the longest established examples of this). Aluminium blades are often constant chord without twist so that extruded sections can be used; sometimes a blade is made up of two or three sections (the leading edge, mid-span and trailing edge) which are extruded individually and fixed together to form the complete blade. Generally speaking this is one of the cheapest ways of producing blades. Examples appearing in the reference list are the troposkien vertical-axis machines by DAF and Dynergy, and the four-bladed machines by Energy Development Co. Steel blades tend in small- and medium-scale turbines to be of rib and spar construction with a thin plate skin, as in the Hermann Brümmer machines; stainless steel is used occasionally. A number of manufacturers use hardened expanded polystyrene or polyurethane as a blade body, with a fibreglass and resin covering for additional strength, waterproofing and smoothness. This can be a particularly cheap method of producing blades once the appropriate tooling is set up and has been successfully adopted by Lubing in its WTG and wind pump production. Fibreglass and plastic composites generally are suited to this application and a number of the manufacturers in the reference list (for example Matsushita and Wind-Matic) have adopted appropriate techniques, sometimes adapted from boat-building practice.

To summarize, many forms of blade construction and material are possible; the best (if there is one) will depend upon many factors, including the manufacturer's own background and expertise, the application and the turbine size.

The issue of pitch variation is probably one of the most controversial in modern WTG design, there being strong advocates of both variable pitch and fixed pitch turbines. For small machines which operate autonomously variable pitch enables overspeeding to be prevented and also provides some measure of power regulation, the latter point being particularly important in connection with battery charging where rapid and frequent fluctuations in electrical output can be damaging to the battery. Dunlite and Aerowatt machines and the horizontal-axis Elektro machines all use variable pitch blades. On the other hand, pitch variation introduces complexity, cost and increased possibility of failure. Fixed pitch is sometimes used on smaller WTGs where overspeed either does not have serious consequences or can be prevented by other means. The Winco machines have fixed pitch blades with a separate centrifugally activated spoiler which opens at high wind speeds, but some field tests have shown that the turbine speed is inherently limited (by its

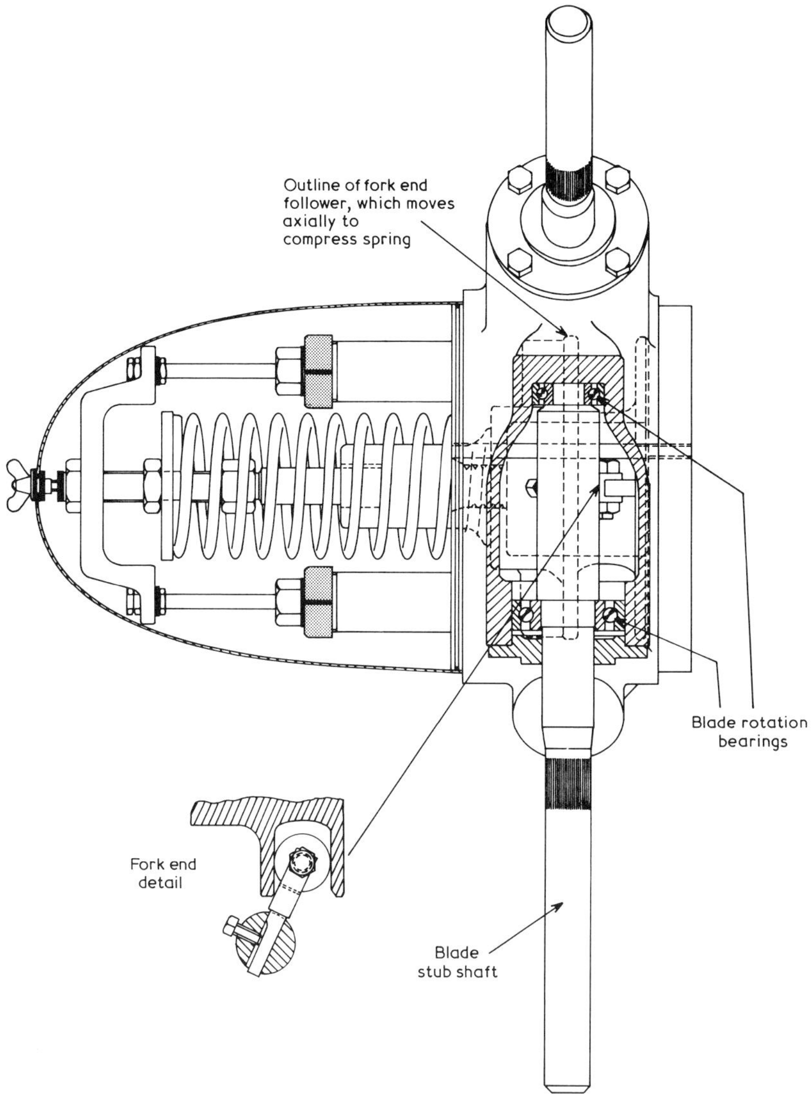

Fig. 3.2 Illustration of centrifugal pitch governing system (by kind permission of Dunlite Electrical Co.).

own drag) even without the spoilers fitted. Other speed regulation techniques such as yaw can also be used in conjunction with fixed pitch turbines to limit rotational speed (yaw being commonly used for this purpose in the traditional multi-blade wind pump).

Pitch change mechanisms in small WTGs tend to be of the centrifugally activated type, using a fly-ball governor principle at the turbine hub. An example of this (showing the hub arrangements of a Dunlite machine) appears in Fig. 3.2. Other means of torque and speed control such as flaps and rotating tips have also been attempted but are generally used more in large (megawatt scale) machines.

A number of machines are designed to operate with the generator directly connected to the 50 Hz or 60 Hz local distribution network. With the use of conventional synchronous or induction generators this results in at least nominally constant speed operation. The turbines of such machines, whether large or small in diameter, can operate satisfactorily with fixed pitch blades, the delivered power being limited by aerodynamic stalling of the blades at high wind speed (the angle of attack then becoming large). Variable pitch blades are used in this application for making the power output more even at high wind speeds, for reducing stresses in high winds and to reduce the rating and therefore cost of the mechanical transmission and generator. However, most machines in the medium size range (up to 20 m diameter say) use fixed pitch blades for simplicity, reliability and cheapness (the machines of Energy Development Co. and Wind-Matic are notable examples of this). Variable speed operation when network coupled can be achieved with variable ratio transmissions, special generator types or rectifier/inverter systems. The advantage is that by allowing the turbine to run at or near its optimum tip speed ratio over a wide wind speed range, the overall operating efficiency can be improved and the delivered energy increased. (This increase may however be offset by the reduced efficiency of the transmission, generator or rectifier/inverter.) It has also been claimed that fixed pitch, fixed speed machines experience considerable fluctuating loads during periods of stalled (high wind speed) operation. The seriousness of this and the extra energy yield from variable speed operation have yet to be clearly demonstrated and it is difficult to make any definitive technical or economic comparison of fixed and variable speed operation in the network connected application. Equation (2.8) and Fig. 2.2 have shown the nature and importance of tip speed ratio in determining turbine output and performance. For low-solidity turbines, maximum power coefficient is frequently achieved at values of λ in the range 6 to 12 and depending upon the design wind speed this results in rotational speeds from about 2000 r.p.m. for turbines of less than 1 m diameter down to less than 50 r.p.m. for megawatt-scale turbines. Low-speed generators suitable for direct coupling at the bottom of this speed range are expensive, very large in diameter, relatively inefficient and difficult to build. There seems little doubt that at this end of the wind turbine range it is preferable to use a speed increasing transmission of some kind between turbine and generator. Generators for operation at more than a few hundred r.p.m. coupled to smaller high-speed turbines are however, quite feasible and are indeed preferable because of the avoidance of a gearbox. Direct-coupled machines are exemplified in the reference list by the small Aerowatt and Sencenbaugh machines. It is clear that a transition point from direct-coupled to speed increasing transmission must exist somewhere in the power and size range; this transition is not defined but probably lies in the 1 to 3 m diameter, 1 to 2 kW rating region. Belts, chains and hydraulics have all been used but the most widely used and probably most successful is the gear transmission in its various forms from helical spur to epicyclic. A novel form of transmission is the rim-belt system made possible by the bicycle-wheel turbine construction of the American Wind Turbine range developed at Oklahoma State University. A problem with this type of device is rim distortion, which tends to throw off the belt during operation. A more sophisticated type of power take-off from the turbine tips involves mounting the rotor of the generator on a rim around the periphery of the blades, the generator stator being mounted around the turbine periphery. This principle has been pursued experimentally but has not reached the commercial stage. It simplifies a number of aspects of the machine construction but calls for the design of an unusual configuration of generator.

In the horizontal-axis class of wind generators, nearly all machines have mechanical transmission and generator housed within the nacelle at the head of the tower. This is inconvenient for maintenance and inspection but it does result in a compact and cheap power train. A notable exception is the Windgen 25, which has a right-angle drive in the nacelle and a vertical generator mounted at ground level within the tower base; this arrangement avoids the necessity for sliprings or a twisting cable to transmit electrical power from the nacelle to the tower. Hydraulic transmission is being used in a megawatt-scale machine presently being developed by Bendix in USA. A three-bladed turbine is used to drive a hydraulic pump in the nacelle, the turbine/nacelle assembly being supported by a tripod which rotates in entirety to follow wind direction. The hydraulic connection is made from the pump in the nacelle to the hydraulic motor/generator which is located at ground level. The nacelle and total tower head weight are relatively low in this design, but the efficiency, reliability and economics of the system remain to be proven.

Generator type tends to be governed by rating and application. For network connected machines the most common is the induction generator, which allows direct connection without the necessity for synchronization or voltage regulation. By virtue of its small (but finite) range of operating speed it also permits some absorption of gust energy as kinetic energy through the turbine inertia. The cage induction machine is also the cheapest, simplest and most rugged type of generator. However, its use can present some problems with magnetizing and starting current and reactive power control, especially on relatively high impedance sections of a supply network. DAF-Indal, Energy Development Co., Kongstedmollen and Wind-Matic are all examples in the reference list of induction generator machines. Synchronous a.c. generators (often called alternators) are generally preferred in machines designed for autonomous use (particularly battery charging), or for network connection through an inverter. Although arrangements can be made to self-excite induction generators using capacitors, it is simpler to draw this excitation current from an external 50 Hz or 60 Hz supply. When this cannot be arranged and independent operation is necessary, a synchronous generator (the excitation for which can be separately arranged) is usually preferred. A brushless a.c. generator of this type with terminal rectification for d.c. output requires less maintenance than the traditional d.c. generator (the need for brushes having been totally eliminated) and nearly the whole of the present range of small WTGs for independent operation uses brushless synchronous machines. The use of permanent magnets in these a.c. generators eliminates the parasitic power for the d.c. rotor field and can result in higher efficiencies; permanent magnet machines are used in a number of wind generators in the reference list.

The use of a horizontal-axis wind turbine necessitates an orientation mechanism to follow changes in wind direction. A number of different techniques have been and continue to be used. The simplest is to allow the turbine to run downwind of the tower, and to construct the blades with a coning in the downwind direction to assist yaw stability. In this way the turbine operation is stable and wind following is automatic. This principle is normally adopted only for relatively small turbines (Lubing and Kedco being examples in the reference list). Lack of understanding of the complex dynamic and structural load behaviour of large free-yaw turbines has discouraged their development in the past. Modern modelling techniques and ingenious design features have, however, recently permitted a more adventurous approach, and free yaw is being attempted at present even up to the largest sizes (about 75 m diameter). Apart from simplifying the means of turbine orientation, downwind operation means that the turbine blades deflect away from the tower under the influence of wind thrust. The turbine hub can then be located closer to the centre line of the tower, this being particularly advantageous where highly flexible blades are used and tip deflections under wind thrust are substantial.

The principal disadvantage of downwind operation is that the blades pass through the disturbed flow behind the tower; this results in an increased cyclic loading on the blades and may accentuate dynamic problems arising from aeroelastic coupling of turbine and tower. A further disadvantage is that the gravitational load of the turbine and the wind thrust add together in producing the tower head bending moment; depending on the precise configuration, this may require stronger nacelle bearing and tower structures than is necessary for upwind operation.

There is no definitive answer at this stage to the upwind/downwind issue and the reference list shows a good selection of both types. The lack of any positive evidence favouring one configuration or the other persists through to larger machines, where past and present prototypes are fairly equally split between upwind and downwind operation.

Upwind machines do, however, require some positive means of turbine orientation. In large machines, the tendency is towards servo-control of the nacelle position using a signal from an auxiliary direction sensor. In small machines simpler techniques such as tail-vanes and fan-tails are used. Tail-vanes are the cheapest and least complicated method of orientation. The tail-vane also enables yaw control (mentioned in the fixed/variable pitch discussion) under storm conditions. If the tail-vane is sprung-hinged to the turbine and the turbine is mounted on the nacelle off its centre line, the turbine thrust at high wind speeds deflects the turbine back at an angle, the tail-vane remaining in line with the wind. An example in the reference list is the Sencenbaugh range. (Other machines such as Elektro use a separate high wind speed sensor coupled to a motorized tail-vane control which turns the vane into the same plane as the turbine, the whole assembly then automatically positioning itself parallel to the wind.) A related technique, although not depending on the use of the tail-vane, is to hinge the turbine such that in high winds the thrust turns the turbine upwards into a horizontal plane, thereby being insensitive to horizontal winds from any direction. The Yamada machine (in the reference list) and North Wind Power machine (in the pre-production list), are examples of this, sometimes known as the 'up' type.

However, since turbine moment of inertia increases roughly as the fifth power of diameter, the tail-vane area necessary to achieve an adequate steering function rises very rapidly with turbine size, soon becoming impractical (the upper limit probably lying in the 10 m to 20 m diameter range, depending upon turbine type). The 18 m Aerowatt machine in the prototype machine list is exceptional for its use of a tail-vane. The fan-tail technique is applicable to larger sizes and derives from traditional tower mill practice going back two hundred years. One or two fan-tails are mounted at the downwind side of the nacelle, their shaft being coupled to the nacelle orientation gear by a high gearing of maybe several thousand to one. As a side-wind strikes the nacelle so the fan-tail(s) turn, the gearing being arranged then to re-position the main turbine into the wind. When the turbine alignment is accurate, the fan-tails experience no side-wind and the precession ceases. Some machines (for example the MAN machine in the prototype list) use a single fan-tail, but others (Wind-Matic and Kongstedmollen in the reference list) use a twin fan-tail to ensure sensitivity to wind from the largest possible sector.

This has been a very condensed and simplified review of the main design alternatives in wind generator design and construction. While these matters are of course interesting to the user, his primary concern is in performance, reliability and economics. The following sections deal with matters relating to these subjects.

3.3 Energy output prediction and performance matching

Unlike diesel generators, which produce a controlled output substantially independent of circumstances and surroundings, WTG output is strongly dependent on natural wind conditions at the selected site.

The performance of a wind turbine generator is characterized principally by its power–wind speed curve, the main features of which are shown in Fig. 3.3. Three points on the curve, the cut-in wind speed, the rated wind speed and the cut-out wind speed are especially important in calculations of machine energy output. There is a common tendency to concentrate upon rated power as the uniquely important parameter, and to use 'load factor' or 'kWh p.a./kW rating' as multipliers which depend upon the average wind speed at the intended site. Such multipliers can be related to the ratio of rated wind speed/site mean wind speed for a crude estimate of energy production, but this relationship is usually derived from a simplified power–wind speed characteristic and a general 'normalized' wind speed–frequency distribution. Errors can be introduced here both in the power–wind speed curve and in the characterization of the site wind regime.

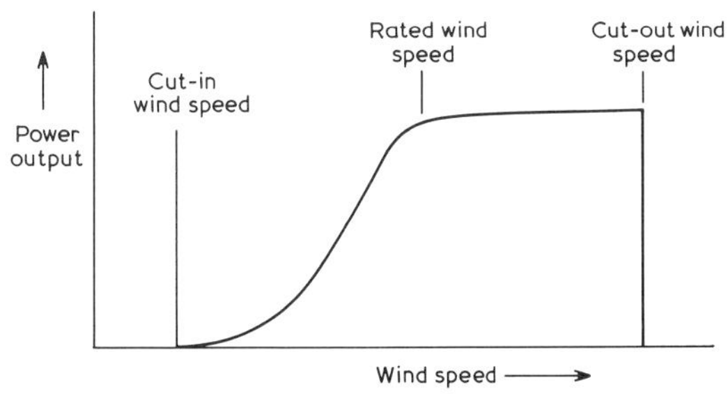

Fig. 3.3 Power–wind speed characteristic for typical plant.

Since the shape and nature of a power–wind speed curve can vary substantially between machine types and the frequency distribution of wind speeds can also vary between sites and areas of the world, it is preferable usually to undertake more specific calculations based on

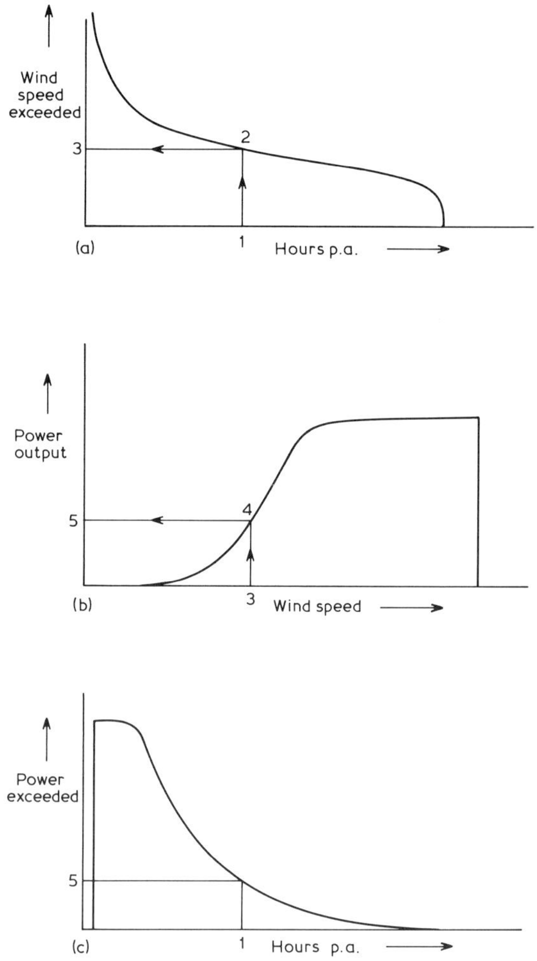

Fig. 3.4 Construction of a power duration curve.

actual characteristics. This can be done with the use of a wind speed exceedence (or wind speed duration) curve for the site, corrected to the hub-height of the machine as explained in Chapter 5. This curve shows the wind speed reached or exceeded for certain fractions of the year. The ordinates of the curve can be translated into

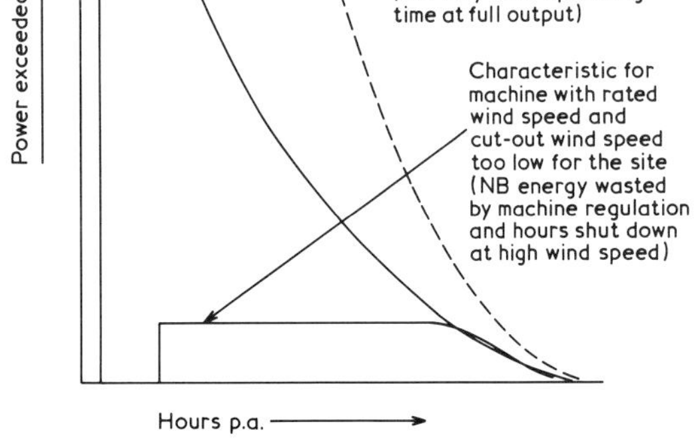

Fig. 3.5 Badly matched power duration curves.

power exceedence using a power–wind speed curve for a particular machine to give a power duration curve, and the area beneath this then represents the annual energy production. Fig. 3.4 shows the process of combining wind speed duration and power–wind speed characteristics in this way.

The power duration curve can assume a wide variety of shapes, depending upon the matching of the machine characteristic and the site wind regime. Examples of 'bad' matching for maximum energy production would be a machine with low rated wind speed at a relatively high wind speed site and the converse (Fig. 3.5).

Guidance on the issue of matching can be derived by comparing the power duration curve with its 'ideal' counterpart, that is a power duration curve calculated using the same wind speed duration curve with a power–wind speed characteristic defined entirely by

$$P = \frac{1}{2}\rho A V^3 \tag{3.1}$$

where A is the same swept area as in the practical wind turbine being evaluated. The latter curve results of course in a substantially higher energy output indicated by the upper (broken) curve in Fig. 3.5. The ratio of energy output predicted for the practical machine to that for the ideal case gives some idea of the loss resulting from the regulation of output above rated wind speed, and also the general inefficiencies and overall operating power coefficient of the machine. Since the machine constant is known as C_p, the power coefficient, it seems appropriate to call this ratio of energy outputs the energy coefficient C_E. Judgments on C_E are best made on a comparative basis to evaluate the 'matching' of several alternative machines at a particular site; it should be possible, however, to achieve values in the range 0.15 to 0.20.

To assist the user in reasonable 'matching' of this kind some manufacturers offer several ratings for individual turbine diameters. Humblot and Aerowatt of France and Kedco of the USA are notable in this respect and the simple summaries of their ranges are shown in Tables 3.1 to 3.3. This approach not only offers the user a range of different power–wind speed profiles in order to attempt optimum matching for the site mean wind speed and energy demand in question, but it also enables the manufacturer to effect some economy in production costs by a rationalization of generator and turbine types. The Kedco range in Table 3.3 is particularly notable in the latter respect.

The wind speed duration curve for a site can be predicted from a knowledge of the mean wind speed (measured for instance by a simple integrating cup-counter anemometer recording run-of-wind) and the shape of the 'regional' duration curve, obtained for instance from recording meteorological stations within the same area. The meteorological station curve ordinates can be divided by the mean wind speed at the station to give a normalized curve, which can then be factored to predict the curve for the prospective wind

Fig. 3.6 Aerial view of Danish test site at Risø, from the west. (Machines are sited by the road running from left to right in the foreground.)

Fig. 3.7 Taketoyo test site, Japan, showing Yamada (foreground) and Fuji wind turbine generators.

Table 3.1 Range of turbine diameters and ratings offered by Eoliennes Humblot (numbers in the table represent rated wind speed in m/s).

Rating (watt)	Diameter (mm) 500	850	1200	2500	3100	4200	4600	5200	8000
25	7								
50		8	7						
100				5/6	4				
300				7					
500					7				
1000						7	6		
2000								7	
3000								8	6
5000									7

Table 3.2 Range of turbine diameters and ratings offered by Aerowatt (numbers in the table represent rated wind speed in m/s).

Rating (watt)	Diameter (mm) 1200	2000	3200	5000	9200
30	7				
50	9				
100			5		
150		7			
300			7		
600			7.5		
960				7	5
1200			10		
4000					7

Table 3.3 Range of turbine diameters and ratings offered by Kedco (numbers in the table represent rated wind speed in m/s).

Rating (watt)	Diameter (mm) 3600	4880
1200	9.3–9.8	7.1–7.6
1900	10.7–11.2	8.9–9.3
2000	11.1–11.6	9.3–9.8

power site. This method is relatively crude and further comment appears in Chapter 5.

A further general point to note concerning the use of wind speed duration curves is that they are generally compiled from mean hourly data. The power–wind speed relationship for any machine is however non-linear (tending to a cube relationship as we have seen) and the wind speed averaging process based on hourly readings can therefore be expected to underestimate the average power production. The ratio of actual energy production to energy production predicted by the method described is known as the energy pattern factor. This factor has been shown by measurement to be as high as 1.1, that is approximately 10 per cent more energy is produced than might be expected by power duration curve prediction. The WTG energy production is affected also by the machine's inertia and response to gusts, its sensitivity and response to changes in direction and the corresponding proportion of operational time spent in a position of yaw. These influences generally tend to reduce WTG output and are discussed in the following section.

3.4 Testing methods and characteristic definition

The engineering purpose of testing can be for proof of integrity and for validation of performance.

At this stage there is no established principle for short-term integrity testing, this being largely dependent on a simulation of maximum gusts and, more difficult, accumulated turbulence representing a lifetime fatigue loading, with progressive degradation or damage due to other environmental effects. Accelerated testing is problematic and it seems at present that the only reliable proof of integrity is long-term field service in a good spectrum of operating environments. This exists of course only for the longer established products such as Dunlite, Elektro and Aerowatt, and even these products do not have unblemished records. It may therefore be some years before recent entries to the business can justifiably claim satisfactory long-term integrity and reliability.

Performance testing is quite different in its nature but nevertheless presents a few problems. The principal requirement is to determine accurately the power–wind speed characteristics of the machine, and this can be attempted by wind tunnel testing, by towing tests or by field testing in natural wind conditions. Few standards or even codes of practice exist at present for the testing of wind generators or the calculation of their power–wind speed characteristic and energy production. Such documents are presently being drafted in the USA and a recommended practice has now been issued under the aegis of the International Energy Agency (IEA) collaboration on wind energy R & D (Frandsen, 1982).

Wind tunnels provide steady and controlled conditions with the availability of generally good instrumentation, these circumstances in principle being ideal for determining the power–wind speed relationship. However, to be fully effective it is important that the percentage blockage that the wind turbine swept area presents in the cross-section of the wind tunnel is fairly small (about 5 per cent) so that the wake can develop without constraint. Full-scale turbine testing becomes more expensive as diameter increases, because increasingly larger-section wind tunnels are necessary, and at some point it becomes impossible with existing facilities. Techniques for accurate simulation of surface roughness, boundary layer profile and turbulence intensity in the wind tunnel have reached a high degree of sophistication. However, it is doubtful whether these techniques are applicable for full-size wind turbine testing in view of the difficulty of reproducing the full-scale turbulence of natural wind. To produce full-scale eddies with length scales approaching 15 m, say, and a high intensity of turbulence requires an extremely large tunnel cross-section, and natural wind testing is for most purposes the only practical solution here.

One of the problems with natural wind testing is that, having erected the machine, a considerable wait is sometimes necessary before the full range of wind speeds occurs so that the power–wind speed curve can be fully determined. Towing tests provide a partial short-term solution. The WTG, with suitable anemometry, power instrumentation and loads, is mounted on a flat trailer and towed for sufficient distance that a steady incident wind appears at the wind turbine and stable power output measurements can be made. The towing is repeated at sufficient mean 'wind speed' points to determine the power–wind speed curve. Airfields have frequently been used for this technique because of the availability of long uninterrupted distances for straight-line towing. The main shortcoming of towing tests is that the turbine does not experience the normal degree of boundary layer turbulence which is introduced into the wind by its passage over the land surface. Under most conditions the speed of the towing vehicle will greatly exceed natural wind speed at the instant of test and it is probable in this case that the turbulence is much lower than in natural wind of equal speed to that simulated by the towing. Towing tests therefore have a serious limitation in the full simulation of natural conditions. They can however be useful in observing the operating behaviour of the wind generator in high wind speeds, particularly with regard to pitch change, deployment of overspeed flaps, operation of furling mechanisms and other safety devices. A few manufacturers (Lubing and Kedco for instance) test production machines in this way on a small flat-bed truck.

In the absence of simulation methods which can be applied with any confidence at present, natural wind testing is therefore probably the only reliable means of performance measurement. The principal advantage is that it allows not only instantaneous (or more precisely short-term) measurements of power and wind speed to be made, so that the power–wind speed curve can be compiled, but it also permits longer-term energy output/mean wind speed relationships to be established. The latter point is very important since natural wind tests take account of all the short-term fluctuations in wind speed and direction and the way in which the wind generator responds to these. For instance, in cases of rapidly rising and falling wind speed, there will be a delay in the turbine's response due to its inertia; this means that the blades are operating for a proportion of the time at non-equilibrium and probably non-optimum conditions. Similarly, the turbine will have a damped response to changes in direction and it will operate in a position of yaw for a proportion of the time; the yaw results in a reduced turbine swept area presented to the wind and changed blade incidence angles, with a consequent reduction in power output.

It is possible that the changing wind conditions have a noticeable effect on energy production; a direct consequence of this would be that different energy production would be expected from two separate sites with differing turbulence intensity, even if the sites exhibited a similar long-term mean wind speed. This is an area which has scarcely been researched at present and no doubt greater understanding will emerge in the future.

The energy production measured from a wind generator operating in a natural environment takes account of these factors and these data are naturally more valuable to the prospective user than theoretical predictions based on ideal operating conditions.

While field testing has these advantages, there are problems in achieving accurate and significant results from the measurements. The first problem is the difficulty of correlating wind speed measurements around a machine with its power output. It is clearly not possible to measure directly the effective wind speed at the turbine, and the anemometer(s) must be sited some finite distance away. The time and space variations in natural wind velocity (coupled with the finite response of the wind generator's power output to wind changes) therefore create a need to correlate the recorded data. This can be achieved using established techniques developed through wind engineering, but the correlation is made more possible and significant if measurements are averaged over a short period, from ten to twenty minutes, and if the anemometers are mounted at or near the hub-height of the test machine, at least two or three turbine diameters away. To avoid interference by the wake of the wind turbine, it is preferable to use several anemometers sited around the machine, and to take data from those in the sector upwind of the turbine during the period of interest. For machines within the size range of the reference section a fifteen-minute averaging period will in most circumstances ensure that error due to difference in the turbine stored kinetic energy at the beginning and end of the period is insignificant compared with the total energy production through the period. For larger machines a longer averaging period may be necessary.

Other corrections should be applied to data derived from field tests, for instance air density corrections allowing for pressure and temperature. Further interpretations should be considered to allow for site-specific wind characteristics, particularly in relation to the surrounding surface roughness and the local 'wind rose' (direction distribution) which will influence the predominant turbulence intensity and vertical wind gradient at the test site. These factors are discussed in the following section and in Chapter 5.

An important part of the characteristic definition of the machine relates to its survival wind speed, this being defined as the maximum wind speed at which the wind generator will remain intact. Despite the length of production and operational experience with some types of wind generator this remains an ill defined parameter. Practical data from failures (even where properly documented) are of little help because the failure point is influenced by the fatigue history of the machine and its production, erection and maintenance detail. Some manufacturers calculate and specify a survival wind speed but this usually applies to smooth undisturbed wind conditions without any allowance for turbulence. The true survival wind speed at a practical site may be lower than this calculated value, but by what extent it is impossible to be certain. This uncertainty is symptomatic of the need for further research into the dynamic loadings on wind turbines under turbulent conditions, so that a proper basis can be developed for prediction of fatigue life and the related issue of survival wind speed.

A further issue which is important in the testing of machines, and the last covered in this brief review, is the nature of the connected electrical load. This is probably the least well defined part of the performance data presently issued by manufacturers, and inadequate information is available to make any coherent or consistent entry concerning this in the reference section. The question of matching of wind turbine output and load characteristic has been touched upon at the end of Chapter 2, and it is clear that the connection of differing loads (that is impedances) to a WTG output terminals will influence the speed and therefore the delivered power. To enable an informed user to correctly match a machine to his pattern of electrical power usage (for maximum energy production), it is important that manufacturer's tests include measurement of power delivery and wind speed for a range of different connected impedances. Some manufacturers (Humblot for instance) offer a speed increasing transmission with the facility for change of ratio; this too should be included as a test variant when putting forward performance data.

3.5 Test facilities

There are now a number of national wind energy R & D programmes, especially in Europe and North America. Some of these concentrate upon the design and development of large megawatt-scale machines, but there is a growing recognition also of the importance of small- and medium-scale WTGs. The most notable of the small/ medium machine programmes is in the USA, where a large test facility has been set up to support the development and commercialization programme. A more modest but nevertheless important test centre has also been established in Denmark and there are emerging facilities in Japan, Sweden, West Germany, Netherlands, China and Canada. The test sites tend to be set up with varying objectives and differing degrees of sophistication, complexity and performance.

The USA test facility at Rocky Flats, Colorado, as described in the *First Annual Report* (1978), is the largest and most sophisticated presently existing. It is located on a 160 hectare level site near Golden, and is operated for the US Department of Energy by Rockwell International. The purpose of the facility is to provide test data (qualitative and quantitative) to prospective users of wind generators and wind pumps, size being limited to a maximum of 100 kW design output. Either commercially available or prototype machines may be tested. It is planned that the site will ultimately be laid out to accommodate 50 machines at any one time, a variety of towers being provided for mounting purposes. Among the machines evaluated or presently under assessment (and appearing in the reference list) are:

- Altos BWP-8B
- Amercian Wind Turbine AWT-16
- Dakota Sun and Wind SI-4
- Dunlite 2 kW
- Elektro WVG 50G
- Enertech 1500
- Kedco 1200
- Millville 10-3-SE
- Pinson Cycloturbine C2E
- Sencenbaugh 1000
- Whirlwind A 240
- Wind Power Systems, Storm Master

Measurements made on the machines include voltage, current, turbine speed, orientation and pitch angle. Wind data are taken at 10 m and 40 m height on 40 m towers, at the top of 15 m portable towers and on the support tower of each machine, and this is correlated with individual machine behaviour. In addition to this long-term performance measurement, intensive monitoring is being carried out on some machines; this includes measurement of machine and tower strains, accelerations, forces and moments, and operating temperatures (*Second Interim Report*, Vol II, 1979). Data collection and analysis is by microprocessors at each site coupled to a central computer. The resulting information is published in semi-annual technical reports, monthly manufacturer data reports, quarterly consumer data reports and final reports for each machine tested. References to the published reports appear in the data sheets (Section 7.1.1), where these have been published.

The Danish test site for small wind generators is located at and run by Risø National Laboratory, Roskilde. The site has been described by Petersen (1980). Planning of the facility began in June 1978 and operation began in May 1979. Machines manufactured in Denmark are tested without charge, the tests on individual machines normally being expected to run for 3 to 6 months. The site is designed to test six machines at one time; it has good exposure to Roskilde Fjord to the west and low undulating land to the east (these being the prevailing wind directions). An overall aerial view of the site from the west is shown in Fig. 3.6. Seven measurements are taken from the meteorological tower at the site and a total of 22 signals can be recorded from each wind generator. Of these 22 signals 3 are constantly connected (for power, r.p.m. and shaft torque), 19 being available for short-period intensive measurements. Data is processed on a central minicomputer. Since the majority of Danish machines are network connected, the facility is set up to facilitate machine operation on the local network. Of the units listed in the reference section, the following were under evaluation in the early stages of the operation of the test site: Poulsen 30 kW, Riisager (now Wind-Matic) 30 kW, Vindkraft 55A15

Fig. 3.8 Part of Annaka test site, Japan, showing Matsushita FY-04 WG1 wind generator.

and SJ.10 kW–R. Of these only the SJ machine is not intended or suitable for network operation. The Danish Ministry of Commerce have been offering prospective users of wind generators a grant on the machine cost and this may be conditional on adequate performance in tests at Risø. The presently agreed budget for operation of the facility is for a three-year period.

The intention of the Japanese test sites is much more to demonstrate applications than to concentrate on detailed performance monitoring. The Science & Technology Agency are responsible for the three sites which are at Taketoyo, Annaka and Kanasawa. The Taketoyo site (Fig. 3.7), which has been testing a Fuji 5 kW and a Yamada Keifu 1, is set up to show the use of wind generators in agriculture (for instance glasshouse heating). This site, being near the south-east coast, has a relatively mild climate all year round. The Annaka site (Fig. 3.8), which has Matsushita FY-04WG1 and Yamada Keifu 1 machines installed, has the intention of demonstrating battery charging for electric vehicles (in this particular case golf carts). Annaka is an inland site with relatively low wind speed and relatively high turbulence. The third site at Kanasawa has an Elektro WV25G and a Yamada Keifu 1 set up to show the use of wind generators for domestic power supply. Kanasawa is on the north-west coast which is prone to harsher winters with snow and ice. Instrumentation at the three sites is relatively simple compared with the US and Danish facilities, and is intended to give indicative guidance on machine performance.

The interesting features of the Japanese test sites are the emphasis on the application, and also the selection of several sites covering a range of wind and climatic conditions.

The West German wind energy programme also includes a test facility for small wind generators, although this is not so far advanced as the US, Danish and Japanese facilities. Up to ten marketable wind machines for different applications in the range of 10 kW will be operated for two years at this facility, which is located on the island of Pellworm in the North Sea. The basic objective is the investigation and comparison of small machines with respect to their reliability and maintenance over long periods. Characteristic measurements will also be made in order to support applications studies. The project is being managed by Gessellschaft für Kernenergieverwertung in Schiffbau und Schifffahrt mbH, and has been described by Fries (1980).

Although Canada has no national wind energy programme at present, there is an interesting Atlantic Wind Test Site being established on Prince Edward Island under the management of the Institute of Man and Resources of PEI. It is planned to set up between eight and twelve test pads initially, with a multi-channel data and control link from each pad to a central data processing facility. Each pad on the 75 hectare site will be instrumented for wind speed, direction and temperature, and the facility for measurements of turbine speed, direction, rate of change of direction, blade pitch position, power output, blade and tower strain on each machine will be available. It is intended that the site be available to Government and private users. Three categories of testing will be offered, concentrating upon proof of concept, performance and reliability; it is expected that the site will be used initially to evaluate machines in the 1 to 50 kW rating range, but WTGs up to 2 MW rating could be accommodated. The facility is at North Cape, which is on the north-western tip of PEI and has a 300° clear exposure to the Gulf of St. Lawrence. Salt-air corrosion, freezing spray and atmospheric icing all occur on this site at various times of the year and the annual mean wind speed (at 10 m height) is about 6.25 m/s. It is especially interesting to note that the detailed wind characteristics at the site were checked by wind tunnel testing of a 1:500 scale model (Irwin and Lodge, 1980).

Other examples of test facilities, some of a more modest or informal nature or at an earlier planning stage do exist in a number of other countries. Also, there are a number of instances where lighthouse authorities or radio/TV/telecommunication organizations have undertaken field testing of commercial machines to enable their own assessments of the relative merits of different machines and whether WTGs as a whole are worth considering for remote power sources.

Overall, it can be expected that during the coming years a significant amount of information will become available either from test organizations or from the manufacturers whose products have been tested. A limiting factor of field testing is always the particular site-dependent wind and environmental conditions to which the machine is exposed. The natural conditions at the test sites described here vary quite widely and it is encouraging to think that within two or three years a broad spread of data will become available representing experience across the climate and wind range.

References

Blackwell, B.F., Sheldahl, R.E. and Feltz, L.V. (1977), *Wind tunnel performance data for two and three-bucket Savonius rotors,* Sandia Laboratories, New Mexico, USA, report No. SAND 76–0131.

Blackwell, B.F., Sullivan, W.N., Reuter, R.C. and Banas, J.F. (1977), *Engineering development status of the Darrieus wind turbine,* Sandia Laboratories, New Mexico, USA, report No 76–0650.

Brulle, R.V. (1977), Giromill wind tunnel test and analysis, *Proc. Third Wind Energy Workshop,* Washington DC, USA.

Clare, R. and Allan, J. (1981), Development of the Musgrove Vertical Axis Wind Turbine, *Proc. Third British Wind Energy Association Conference,* Cranfield, UK.

Frandsen, S., Trenka, A.R. and Pedersen, B.M. (1982), *Recommended practices for wind turbine testing.* Part 1. *Power performance testing.* Report submitted to the Executive Committee of the IEA Programme for R and D on Wind Energy Conversion Systems.

Fries, S., Borchers, D. and Petersen, G. (1980) *Comparative investigations on the performance of small wind energy conversion systems,* GKSS 80/E/54.

Grylls, W., Dale, B. and Sarre, P.-E. (1978), A theoretical and experimental investigation into the variable pitch vertical axis wind turbine, *Proc. 2nd International Symposium on Wind Energy Systems,* BHRA Fluid Engineering, Cranfield, Bedford, UK, Volume 2, pp. E9–101 to E9–118.

Irwin, H.P. and Lodge, M.A. (1980), *Wind speed distribution over a 1:500 scale model of the Atlantic Wind Test Site, North Cape, Prince Edward Island,* Institute of Man and Resources, Prince Edward Island, Canada (21 March).

Musgrove, P.J. and May, I.D. (1978), Development of the variable geometry vertical axis windmill. *Proc. 2nd International Symposium on Wind Energy Systems,* BHRA Fluid Engineering, Cranfield, Bedford, UK, Volume 1, pp. E4–39 to E4–60.

Petersen, H. (1980), *The test plant for and a survey of small Danish windmills,* RISØ–M–2193, Risø National Laboratory, DK–4000 Roskilde, Denmark.

Rocky Flats Small Wind Systems Test Center Activities (1978), *First Semi-annual Report,* Volume I, Description of the National Small Wind Systems Center, RFP–2920/3533/78/6–1 (28 September).

Rocky Flats Small Wind Systems Test Center Activities (1979), *Second Interim Report,* Volume II, Controlled Velocity, Vibration and Dynamometer testing of Small Wind Energy Conversion Systems, RFP–3004/3533/79/7–2 (July).

Templin, R.J. (1974), Aerodynamic performance theory for the NRC vertical-axis wind turbine, National Research Council, Ottawa, Canada, report No. LTR–LA–160.

4 Wind Pumps

The majority of attention in this book is devoted to wind turbine generators and this reflects the substantial and increasing interest in this field and the diverse design methods and technology being used. The use of wind power for pumping is, however, a much longer established application, probably going back several thousand years in one form or another.

Pumping of water has proved an ideal application for wind power because of the ease of effectively storing potential energy in raised water. This overcomes the intermittency of wind in a simple and relatively cheap fashion that is not available for wind generated electricity.

4.1 The principal applications

There are three discernible types of application. The first is high-lift pumping (often from a borehole), where the water has to be raised say 10 m or more. The second is low-lift pumping, usually for drainage or irrigation purposes, and the third is strictly not pumping but the related application of pond aeration.

It is the high-lift pumping application which has encouraged the development of the high-solidity turbine with a large number of broad metal blades. This type (often called the 'American' type) was referred to briefly in Section 2.2 and it originated over a century ago in the USA following a substantial need for pumped water in the farms and ranches of the mid west. Similar machines were developed against the same need in Australia, South Africa and Argentina. The predominant need for the machines has been to provide watering for stock, and it is said that about a million units have been installed world wide. An arbitrary lower level of 10 m lift has been mentioned to distinguish between high-lift and low-lift irrigation; at the deep end of the range some of the largest wind pumps are now capable of operating at a maximum lift of about 400 m.

The lower-lift machine for drainage seems to have evolved against a more local requirement to service individual fields or small areas. The machines tend to be smaller and simpler and have probably not been produced in such large quantities. The market also tends to be more localized, focussing upon north-west Europe, particularly the Netherlands, Germany and Denmark where the land is flat and the climate quite wet. The differences in design adopted here compared with the high-lift application are discussed in the following section.

Pond aeration is a specific application which is only loosely connected with pumping, but is included here for completeness. The equipment has been developed for use in static water having a tendency to stagnate, and comprises an underwater propeller or fan-type device driven directly by the wind turbine. Use of these machines also keeps a limited surface of water clear of ice in the winter and one manufacturer claims that his aerators reduce crushing damage to boat docks, prevent ice heaving and damage to beaches and provide clear drinking space for livestock.

4.2 Main design features

The design of the conventional style of wind pump (for the high-lift application just described) has remained largely unchanged for over fifty years. The turbine comprises a number of curved sheet metal blades which increase in width going outwards from the centre. The blades are not articulated and are fixed at their inner end to a circular rim. A further rim is situated near to the outer edge of the blades, running around each blade and providing it with radial and peripheral support. The rims, which are usually segmented, are mounted from the hub by a number of radial arms, each usually having two legs, one sloping forward and the other backward from the turbine plane, to spiders on the turbine shaft.

Fig. 4.1 indicates the layout of a Dempster machine, showing traditional engineering terms for the various parts.

For larger machines there are often several rims at various radial positions on the blade and a larger number of radial arms. The brake for the turbine is often incorporated with the rear of the two spiders, and is usually either of the brake band or internal expanding type. The turbine is overhung on the shaft from the transmission casing and the front bearing is one of the important features on which manufacturers compete. Wooden sleeve, cast white metal, cast iron or self-lubricating ball-bearings are all used, each with an individual claim for superiority.

The transmission itself usually comprises a divided drive speed reducing gear with a ratio approximately 3:1. The divided drive (with the two large gears either side of the pump rod) is claimed to distribute the load and increase working life (Fig. 4.2). Wakes and Lamb use only a single drive train and the large (No. 18) Climax machine is direct coupled, without any speed reducing gears (Fig. 4.3).

The turbine rotation is converted to reciprocating motion in the pump rod through a crank and connecting rod (or Pitman). This is housed within a casing which constitutes the main-frame of the tower head unit and is often known as 'the engine'. The rotor, tail-vane and turn-table bearing are all mounted on this main-frame. Casing design has evolved over the years to a point where oil leakage is minimized, the claim for annual lubrication being an important feature.

The tail-vane is a simple fabricated structure whose area and length have again been subject to an evolutionary process. Automatic 'furling' in high wind speed is achieved (as mentioned in Section 3.2.2 for wind turbine generators) by offsetting the turbine axis from the axis of rotation on the tower so that the turbine thrust at high wind speeds tends to slew it round, the tail-vane remaining in line with the wind. The turbine thrust is usually countered by a spring to produce a progressive action. Fig. 4.4 shows a typical layout.

The pump rod connecting from the tower-top unit to the piston at the bottom of the bore-hole comprises three sections. The first, often known as the 'red-rod', runs down to the base of the tower. This is usually of soft wood, deliberately the weakest link in the overall pump rod because it is most accessible for repair. The red-rod connects usually to a brass rod which passes through the 'packerhead', that is the sealing unit for the top of the bore-hole (see Fig. 4.5). The packerhead, which is normally at ground level, also facilitates inclusion of a non-return valve so that the delivery pipe can be connected to a storage tank some height above ground level

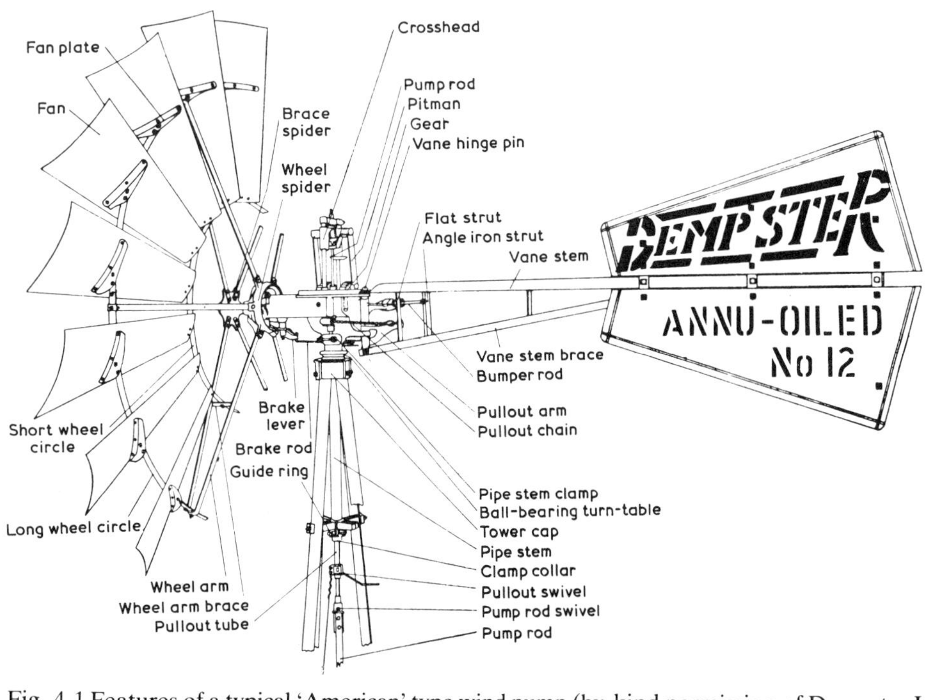

Fig. 4.1 Features of a typical 'American' type wind pump (by kind permission of Dempster Industries Inc.).

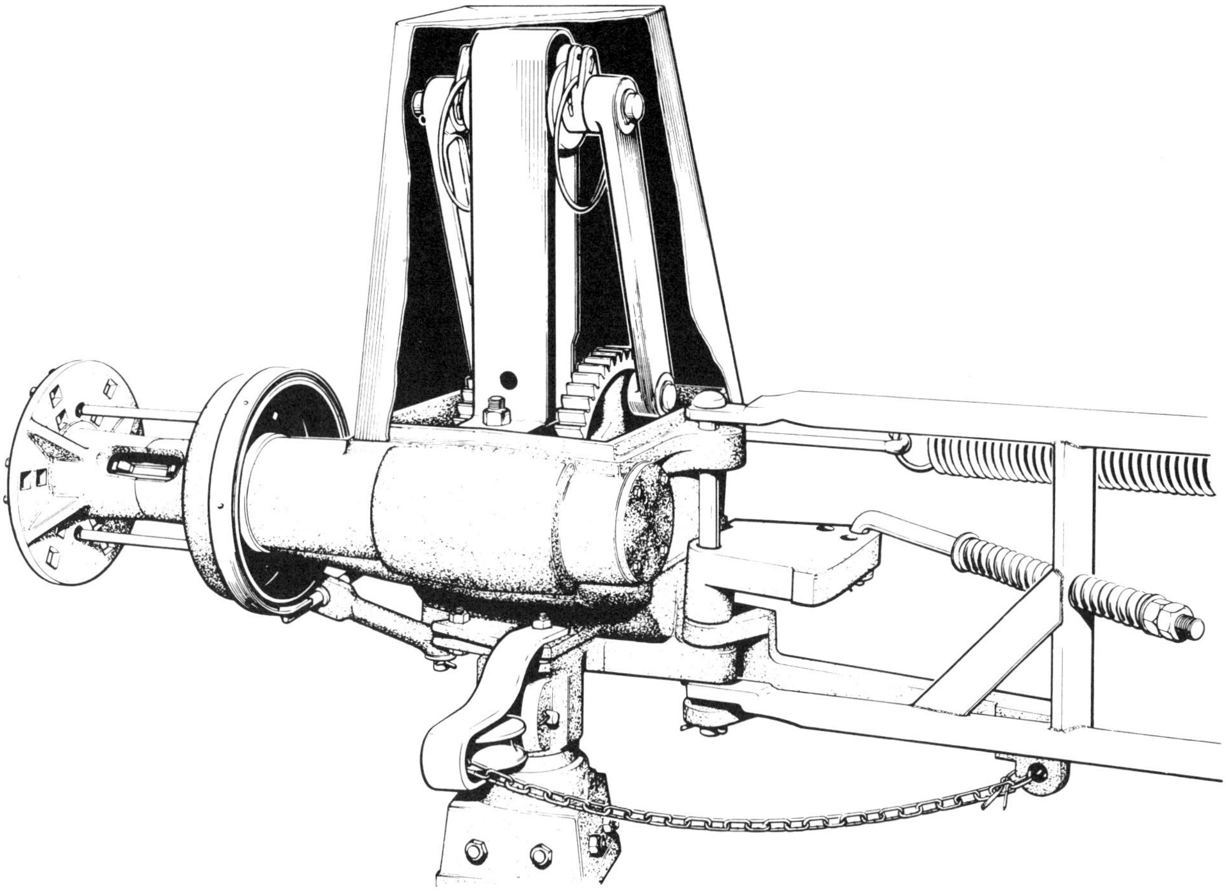

Fig. 4.2 Transmission for an 'American' type wind pump incorporating a divided drive (by kind permission of Climax Windmills (Pty) Ltd.).

Fig. 4.3 Transmission assembly for an 'American' type wind pump without gears (by kind permission of Climax Windmills (Pty) Ltd.).

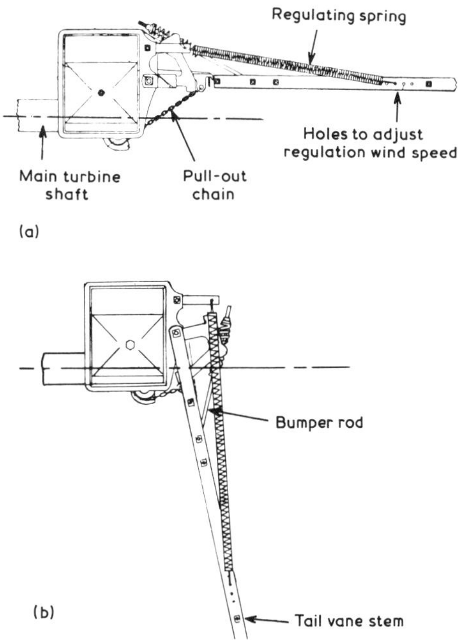

Fig. 4.4 Typical layout of the furling arrangement for a tail vane (by kind permission of Dempster Industries Inc.). (a) Low wind speed position, (b) high wind speed position (vane angle exaggerated).

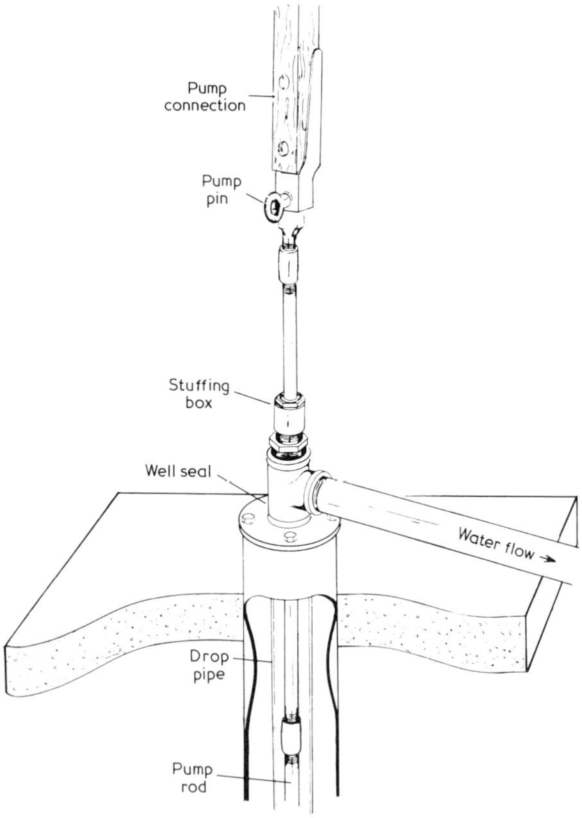

Fig. 4.5 Typical well-head layout (by kind permission of Dempster Industries Inc.).

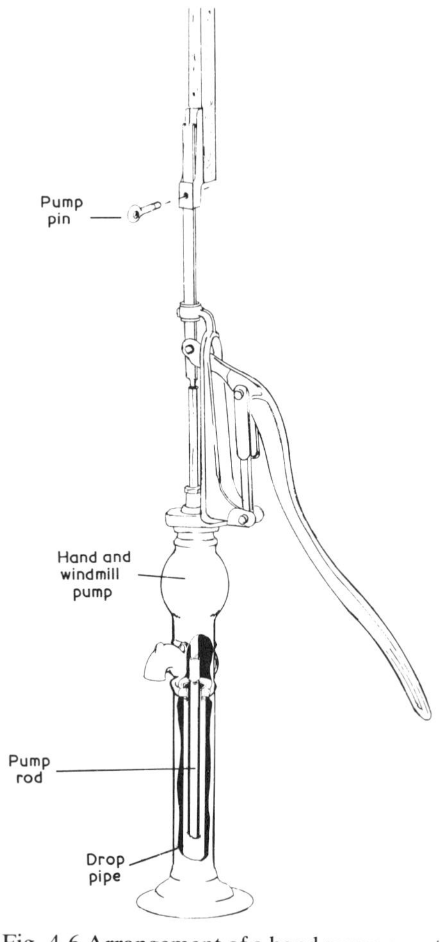

Fig. 4.6 Arrangement of a hand pump over the bore hole (by kind permission of Dempster Industries Inc.).

without risk of drainage back into the well. Some wind pump manufacturers offer a hand pump (Fig. 4.6) that can be mounted on the packerhead so that water can be pumped manually during long windless periods. It is usually recommended that the hand pump fitment is removed before wind pumping is resumed, to avoid mismatching of the strokes and breakage of the red-rod. The section of the rod passing through the packerhead operates in water and air, and brass is chosen to minimize corrosion. The cross-section of the pump rod within the borehole depends upon depth. Shallow pumps use a relatively thick (50 mm diameter) solid steel rod to minimize buckling in the downstroke; the extra weight also assists against resistance from the piston leathers. Deeper holes require a slightly lighter type of rod, usually of hollow steel closed at both ends to provide buoyancy in the water and reduce the weight to be lifted. The deepest holes call for wooden rods, which reduce the overall weight to a minimum. At the bottom of the bore-hole the piston operates in a cylinder (Fig. 4.7) which screws into the bore-hole liner; this liner therefore affords limited clearance to enable the piston to be lowered during erection.

The tower is usually of conventional lattice steel or truss construction; it can be erected either *in situ* or horizontally and raised into position with a gin-pole. For deep wells it is important to have a reasonable height of tower to minimize the number of pump rod sections, since these have to be connected one by one (within the tower) as the piston is lowered down the bore-hole.

More recent years have seen a departure from this traditional style of wind pump in new developments, and the Lubing, Bowjon and Bosman machines are a few examples of this departure. It is not easy to generalize on

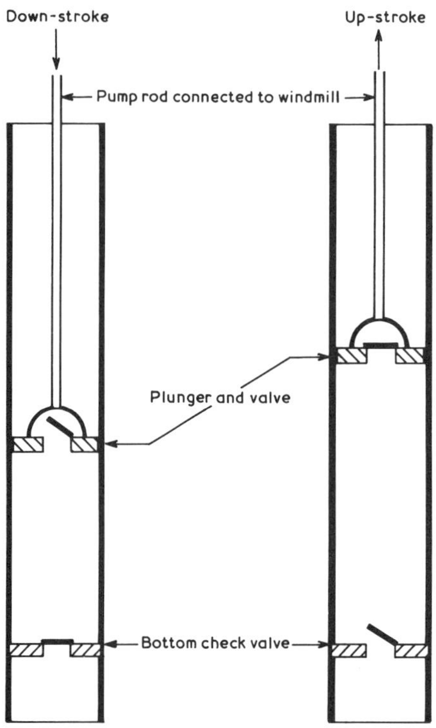

Fig. 4.7 Layout of the pump in the bore hole.

the design of these more modern machines but the following comments are offered.

The overall tendency is a reduction in the number of blades and a lower turbine solidity, resulting in higher rotational speeds and probably a higher power coefficient (especially when aerofoil blades are used, as in the Lubing machines). In some cases this means that the turbine is equally applicable to the wind generation duty, and Lubing for instance use a similar turbine throughout their range of wind pumps and wind generators. Many of the smaller modern European wind pumps tend to favour broader (non-aerofoil shaped) metal blades similar to the American wind pump style, using, however, only 2, 4 or 6 blades. These units are often intended for low-lift operation and the low turbine solidity arising from fewer blades has probably been favoured because of the much lower starting torque required in this application compared with the high-lift duties required of the traditional American wind pump.

A further difference in some modern designs is the use of guyed tubular steel towers, the pump rod being enclosed within the tower tube. For low-lift suction pump applications the pump (often of diaphragm type) is mounted at the base of the tower with a direct coupling to the pump rod (shown in Fig. 4.8). The use of suction

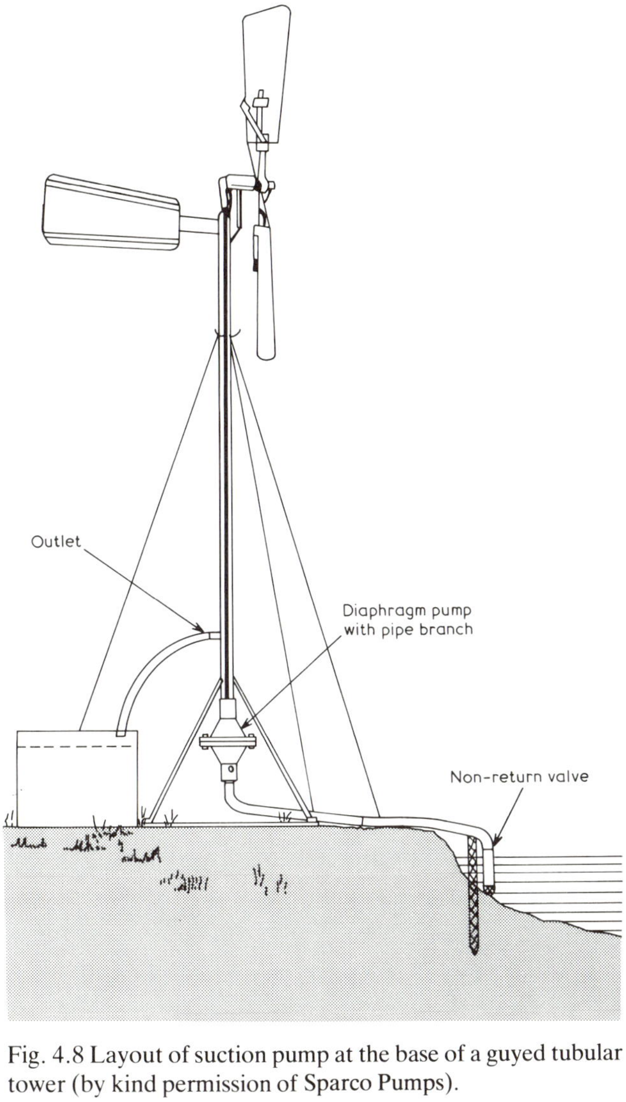

Fig. 4.8 Layout of suction pump at the base of a guyed tubular tower (by kind permission of Sparco Pumps).

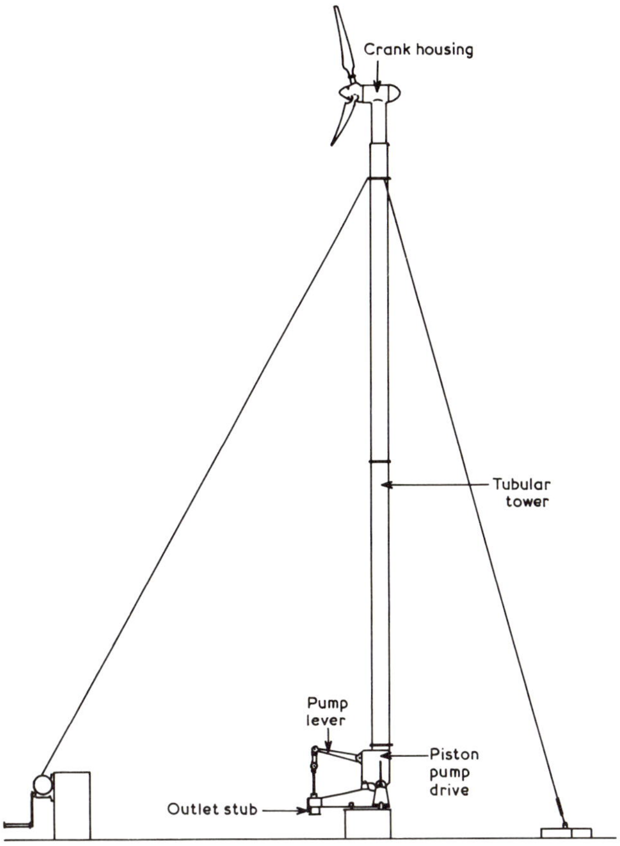

Fig. 4.9 Layout of pump rod in tubular tower, with crank arrangement at tower base (by kind permission of Lubing Maschinenfabrik).

pumps generally has meant that the wind pump unit does not have to be sited over the source of the water (Fig. 4.8). The lift using a suction pump is generally limited to around 6 to 7 m. For high-lift applications the tower tube can of course be located directly over the bore-hole. Lubing in particular offer towers which can be erected and lowered (by one man) about a pivot at the tower base; to facilitate this the pump rod in one machine of the Lubing range operates through a crank at the tower base, the lines of the tower tube and bore-hole being slightly offset (Fig. 4.9).

New materials are being adopted to a limited extent in 'modern' wind pumps. Lubing machines are a leading example, using moulded expanded polystyrene filler with a glass-fibre skin (as described in Section 3.2.2) for the blades, stainless steel for piston rods and plastics for the pump guides.

An interesting design variant which has become commercially available in the last few years is the wind compressor, coupled pneumatically to a Venturi pump. Conceptually the equipment comprises a wind turbine, which can be of normal horizontal-axis type, coupled mechanically to an air compressor taking its inlet from the atmosphere. The output from the compressor is taken underground to a Venturi pump at the bottom of the bore-hole, under a minimum level of submergence. The system has several advantages. The wind turbine and tower can be conveniently located away from the bore-hole, so that best local wind conditions can be utilized. Several wind compressors can be coupled to the same bore-hole. There are no moving parts at the bottom of the well, and also the pump can run dry, accepting silt, sludge or sand without the damage that might result in a conventional piston pump. Three manufacturers (Bowjon, Millville and Windpumpen Zentrale) are currently offering wind compressors and this is one of the most interesting developments in wind pumping for many years, but it remains to be seen whether experience will vindicate the theoretical advantages that are apparent.

4.3 Selection methods and performance matching

Wind pump performance and capability is traditionally and most commonly described using a pumping capacity table. This is the presentation which has been adopted in the reference section dealing with commercial wind pumps.

The starting point for the prospective user is to determine the pumping requirement in terms of pumping capacity or volumetric rate (cubic metres per hour, or, in imperial units, gallons per hour) and the lift or head through which the water must be pumped. These being settled, the pumping capacity table sets out three parameters (turbine diameter, pump stroke and cylinder diameter) which can be combined to give various lift/pumping capacity options. From these options the user can select the nearest to his requirements. For a given turbine diameter and pump stroke, the influence of increasing cylinder diameter is to reduce the lift capability and increase the pumping capacity. The general advice given by wind pump manufacturers is to avoid oversizing the cylinder diameter for a given turbine diameter (the temptation being to achieve greater pumping capacity); this oversizing can result in fewer running hours because of the higher starting torque, and the overall effect is not necessarily to increase overall volume of water pumped. A similar comment applies to pump stroke; the effect of increasing pump stroke (for a given turbine diameter and cylinder diameter) is again to reduce lift and increase pumping capacity. Some manufacturers' ranges do not give any choice of pump stroke, once turbine diameter is selected; where the option exists, however, it is relatively easy to make the change in service, by re-locating the connecting rod in an alternative hole at a different radius in the driven gear wheel. To change cylinder diameter once the pump is commissioned is not usually very practical.

Circumstances can exist where the water storage tank fed by the wind pump must be located at a substantial

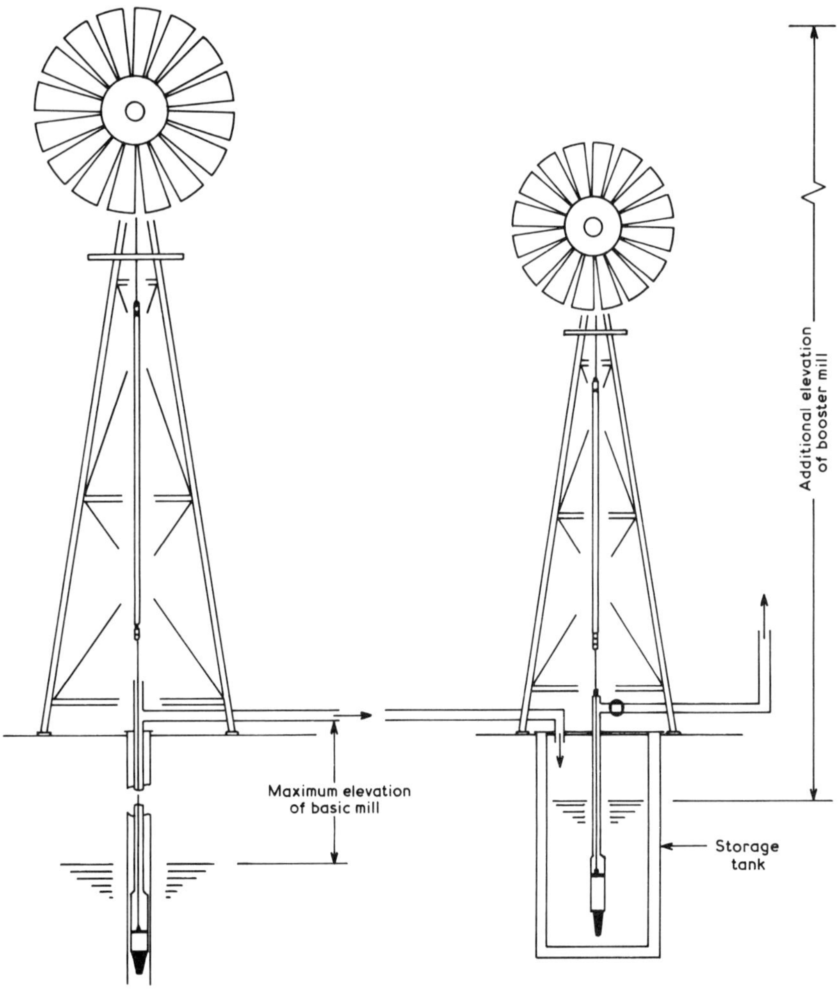

Fig. 4.10 Use of a 'booster' wind pump.

height above ground level at the pump, in order to provide adequate pressure in the water distribution system. This can be achieved either by a tall tower near the wind pump (the tank is mounted integrally in the wind pump tower in some cases, for instance Reymill in the reference section) or by siting the storage vessel on nearby hills. The increased lift can in these circumstances be provided by a 'booster' wind pump. This is especially useful if the additional lift becomes necessary through some change in need after the first wind pump has been erected. The outline of the booster scheme is shown in Fig. 4.10. The delivery from the first pump is fed to a ground level storage tank situated below the booster pump, which then provides the necessary extra lift to discharge the water in the elevated storage tank. It is important when adopting this technique to make sure first of all that the volumetric rate achieved by the two pumps is closely similar (otherwise the ground storage tank will tend to overflow or run dry) and secondly to situate the two wind pumps reasonably well apart. Interference between adjacent wind turbines is at present the subject of considerable research and no clear guidelines yet exist but the general consensus appears to recommend a minimum separation of between five and ten turbine diameters.

A word of caution should be offered concerning the dependence of pumping capacities upon the mean wind speed at the site. Most of the machines included in the reference section incorporate positive displacement pumps, which deliver a constant pressure and will therefore work through a constant lift whatever the operating speed. Consequently the mean wind speed at the site will not affect the machine's capability to work against the required lift from the bore-hole. The wind speed, since it affects the torque and speed developed by the turbine, will, however, determine the volumetric rate. The data supplied by manufacturers concerning dependence upon wind speed are not so explicit and detailed as they are for wind generators. Most pumping capacity tables are accompanied by a note that the figures apply at a certain wind speed, but it is often not clear whether this signifies a short-term constant level of wind speed, or a long-term average wind speed (the pumping capacities then also being long-term averages, of course). Some manufacturers indicate the percentage changes in pumping capacity that can be expected at one or two different wind speeds.

The question of load matching has been touched on in Section 2.4. It seems clear that in general the American-style wind pump does not achieve ideal matching of the turbine characteristics. Dixon (1979) claims that the overall operating power coefficient (equivalent to the energy coefficient defined in Section 3.3) is typically as low as 0.05, this being due to three separate factors.

Apart from the power coefficient of the wind turbine itself (judged overall to be about 0.3) Dixon estimates that the efficiency of turbine to pump matching across the speed range is about 40 per cent, and a further 40 per cent efficiency is imposed as a result of cyclic torque variations required by the pump. The positive displacement pump used in most wind pumps requires a power proportional to speed, but the turbine itself (if operating at constant tip speed ratio) produces power proportional to the cube of speed. Referring back to Fig. 2.9, in which the matching concept is expressed in torque–speed curves, this means that the turbine and pump are operating in a matched condition at only one speed, and throughout the remainder of the operating range the combination must operate in a condition which becomes increasingly inefficient as this point is departed from. A further mismatch of torque–speed curves inevitably occurs through each pump cycle because of the cyclic position-dependent nature of torque demand from the pump and the relatively even level of torque delivered by the turbine. A number of solutions such as continuously variable pump stroke (related to turbine speed) have been postulated, but to the author's knowledge few if any of these have been adopted commercially. It has even been suggested that holes in the pump piston head may improve overall pumping capacity by de-sensitizing the matching problem; there is even some evidence to show that wind pump delivery increases during the early period of service from new, while the leathers become worn.

4.4 Test facilities

In contrast to the intermittent but generally increasing interest and activity in performance testing of small wind generators throughout the twentieth century, very little has been done in connection with wind pump performance testing. This is paradoxical in view of the very much larger historical and current market for wind pumps, but perhaps understandable in view of the relative tolerance of the agricultural market and the application.

It is worth mentioning, however, that at least two of the wind generator test sites described in Section 3.5 have made provision for wind pump testing. A site has been made available for wind pump testing at the Rocky Flats test site in the USA, and also a vertical-axis turbine experimental wind pump was erected at the Taketoyo (agricultural application) test site in Japan.

Reference

Dixon, J.C. (1979), Improving the mechanical load matching of wind energy converters, *Proceedings of the 1st British Wind Energy Association Workshop,* Cranfield, UK.

5 Siting of Wind Machines

It has already been observed in Chapter 3 that the annual output of a wind machine is strongly influenced by the mean wind speed of the selected site, and this in turn critically affects the economics of the installation. It is important, therefore, to be able to predict wind speeds at any site with some degree of confidence, and without recourse to extensive and protracted anemometry which can be both expensive and time consuming.

Apart from the need to select a site with a high mean wind speed, there is also a need to avoid turbulence and local wind field peculiarities, which can place excessive cyclic and extreme loads on wind machines.

The understanding and means of analysis of boundary layer winds and the manner in which they are influenced by topography and local irregularities are, however, not fully developed at present. Predictive techniques therefore tend to be a combination of theoretical approaches, many with varying simplifications of the Navier–Stokes equations, and empirical methods based on measured data from a variety of circumstances, deduction, intuition and local experience. Treated in full, this is a highly complex subject requiring a good understanding of boundary layer theory and meteorology; this handbook attempts only to set out general guidelines for users with the task of selecting a site within a fairly restricted land area to serve his needs.

Although anemometry can be expensive, it is nevertheless a very valuable aid to site selection, and where large wind machines are to be installed it is a particularly important tool, even if only when used in a confirmatory fashion. The final part of this chapter deals with types of anemometers and wind measurement in general.

5.1 Factors influencing wind characteristics at a site

The overall pattern of wind circulation around the earth is shown simply in Fig. 5.1. Broadly there are three main classifications of surface wind. Winds at higher latitudes are of a prevailing westerly nature and have generally highest mean values, particularly over the sea and where they are incident upon coasts after a long fetch over clear ocean. In mid latitudes (say between 15° and 40°) the regimes over the ocean are characterized by much steadier winds with lower mean values, known as 'trade winds'. Near the equator mean wind speeds are generally very low, large ocean areas being commonly

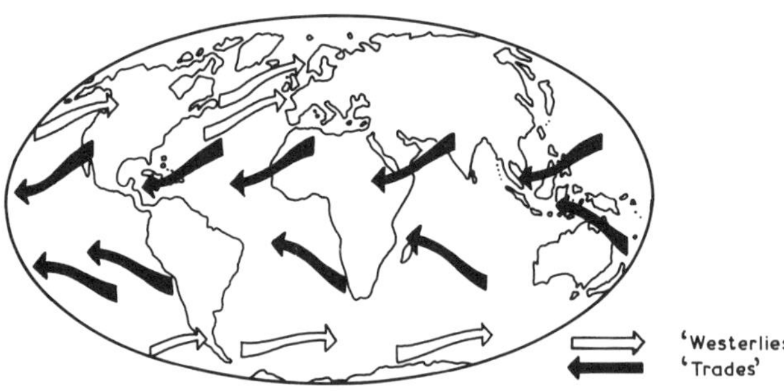

Fig. 5.1 General pattern of wind circulation around the earth.

known as the 'doldrums'. In this broad respect, wind energy and solar energy are, in their general geographical distribution, complementary.

Superimposed upon this global pattern of winds are characteristics of the wind which have been created by local conditions. These locally induced effects can be divided into two types. In areas of relatively high wind speed such as higher-latitude coastal areas or islands in the trade winds, topography is very important in creating local acceleration effects and considerable enhancement of wind speed above the surrounding area can be found at the top of hills and ridges, for instance. In low wind speed areas at the centre of continents or near the equator local thermodynamically induced effects such as valley winds and sea breezes can be much more important, since their magnitude is often greater than the mean'prevailing' or synoptic wind.

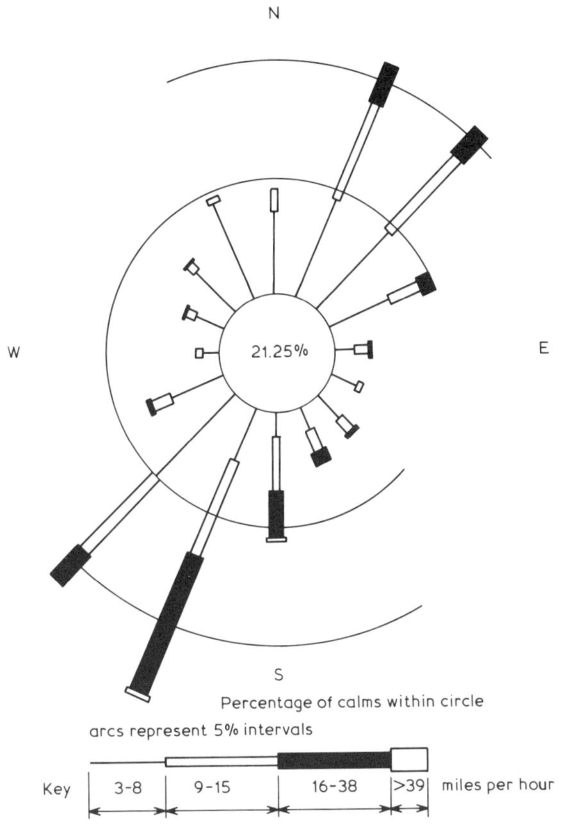

Fig. 5.2 Example of a wind rose (Durban).

In the former of these two types, where synoptic winds dominate, one of the first steps in selecting a site should be to examine the regional distribution of wind direction. This is commonly represented by a 'wind rose' (Fig. 5.2 shows an example) which can usually be drawn using data from the nearest meteorological station. Usually the wind rose will show one or two prominent directions. Ideal sites in that region should have a long stretch of level smooth terrain or sea towards the most prominent direction indicated by the wind rose. If a single wind direction is dominant then it has been pointed out by several workers in the field, notably Golding (1955) and more recently Wegley, Ramsdell, Orgill and Drake (1980), that the best topography for accelerating the flow is a ridge perpendicular to the wind direction; this does not permit 'leakage' of wind around the sides of the contours in the same way as might an isolated hill. In such a case, however, the acceleration or 'speed-up' effect at the top of the ridge decreases substantially when the wind is not in the prevailing direction and the full value of using the ridge as a site will depend upon the proportion of total incident energy which is contained in winds not in the prevailing direction. In areas where the directional frequency of winds is distributed more evenly, hills are preferable as sites. Provided a hill has suitable profile, surface and surrounding environment it can provide substantial speed-up effects near the ground for a range of wind directions, the overall average speed-up being potentially better than for a ridge being subjected to the same range of directions. Measurements and surveys reported by Tagg (1957) and Frenkiel (1961) and subsequent experience has led to an acceptance of the following desirable characteristics in the hill to be used for wind power:

(a) Open exposure in all directions. Similar or higher hills should be as remote as possible, particularly in dominant wind directions;

(b) The best approximation to conical shape;

(c) Smooth upper slopes without trees or any substantial vegetation;

(d) A fairly steep final approach to the summit (the optimum slope is considered to be about 1 in 3.5 or 16°).

Flat topped hills (or ridges) are not particularly satisfactory because of the possibility of stagnation or detachment of the flow, resulting in low mean wind speeds and extra turbulence at the summit.

The height of the hill or ridge is not (within normal limits) of primary importance in affecting the speed-up. Increased altitude does, however, result in reduced power output because of the associated reduction in density (Table 5.1 shows the reductions in temperaure, pressure and density with height as defined in the International Standard Atmosphere).

A further consequence of increased altitude, especially in temperate climates, is the greater possibility or increased occurrence of icing. Glaze icing is probably the most serious type. It is formed by rain freezing upon impact and, because of its greater adhesion, mechanical strength and density is more likely to cause damage through imbalance or high thrusts than rime, which is usually formed under calm conditions, has little mechanical strength and tends to fall away under vibration or rotation. Statistics upon the likely extent and rate of formation of glaze ice are unfortunately not readily available, partly because of its localized nature and the rapidity of accumulation and subsequent disappearance under various weather conditions. (It is worth noting for wind power applications that glaze ice is often accompanied by quite high wind speeds.) Some manufacturers offer blades with special coatings such as Teflon which are designed to shed ice as it is formed; these have been highlighted in the reference section and should be given special consideration if a serious icing

Table 5.1 Variation of air temperature, pressure and density with altitude, as defined in the International Standard Atmosphere.

Altitude (m)	Temperature *	Pressure †	Density ‡
0	1	1	1
250	0.9944	0.9707	0.9762
500	0.9887	0.9421	0.9528
750	0.9831	0.9142	0.9299
1000	0.9774	0.8869	0.9074
1250	0.9718	0.8604	0.8853
1500	0.9661	0.8344	0.8637
1750	0.9605	0.8091	0.8424
2000	0.9549	0.7845	0.8215
2250	0.9492	0.7604	0.8011
2500	0.9436	0.7369	0.7810
2750	0.9379	0.7141	0.7613
3000	0.9323	0.6918	0.7420
3250	0.9266	0.6701	0.7231
3500	0.9210	0.6489	0.7045
3750	0.9154	0.6283	0.6863
4000	0.9097	0.6082	0.6685
4250	0.9041	0.5886	0.6511
4500	0.8984	0.5696	0.6339
4750	0.8928	0.5510	0.6172
5000	0.8872	0.5329	0.6007
5250	0.8815	0.5154	0.5846
5500	0.8795	0.4983	0.5689
5750	0.8702	0.4816	0.5534
6000	0.8646	0.4654	0.5383
6250	0.8589	0.4497	0.5235
6500	0.8533	0.4344	0.5091
6750	0.8477	0.4195	0.4949
7000	0.8420	0.4050	0.4810
7250	0.8364	0.3910	0.4674
7500	0.8307	0.3773	0.4542
7750	0.8251	0.3640	0.4412

* Unity represents 15°C (or 288.16 K) at sea level.
† Unity represents 101 325 N/m² at sea level.
‡ Unity represents 1.2256 kg/m³ at sea level.

possibility is judged to exist. One or two manufacturers go to the trouble of indicating that their turbine will operate and survive with a certain thickness of ice accumulation on one blade with no build-up on the other(s); this situation can arise with partial ice shedding.

Special topographical features other than hills or ridges can sometimes be exploited for wind power. Notable among these features are valleys aligned in the prevailing wind direction, especially where these have local constrictions which accelerate the flow, and 'saddles' or passes between hills. Whilst offering quite interesting speed-up factors in the prevailing wind condition, this type of site can often experience calm or (equally detrimental) turbulent conditions when the wind is in other directions. Such sites should therefore be considered with caution.

In the second class of local conditions, where winds are principally influenced by thermodynamic effects, mountain/valley and sea/land breezes are most noteworthy. Both depend upon local surface temperature differences which fluctuate and/or reverse on a diurnal basis. The winds tend to be weaker than in the synoptically dominated systems previously described but in certain regions they are the only significant source of wind power. In a mountain/valley system the wind flows up a valley in the direction of the mountain during the noon to sunset period, this system being driven by greater surface heating in the mountain region. In the evening to sunrise period the reverse applies, with cold dense air flowing down the valley from the mountain. In the former regime (valley winds) the wind is usually weaker and more variable than in the latter (mountain winds). A rough schematic is shown in Fig. 5.3. The effect is maximized near the mouth of the valley, at the confluence of two valleys or at a valley constriction.

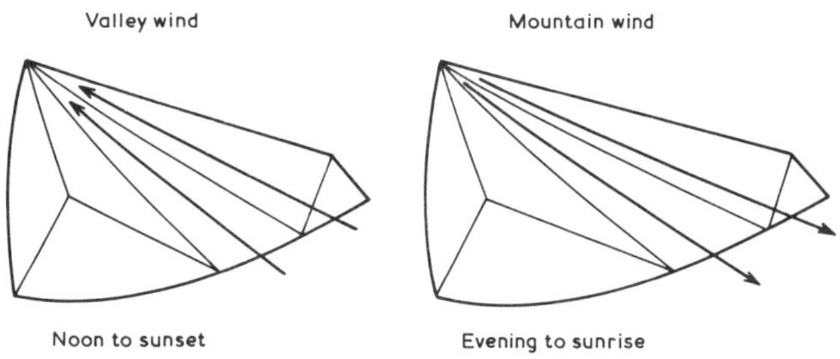

Fig. 5.3 Illustration of mountain and valley winds.

Land/sea breezes are most prominent on the west coasts of continents in subtropical areas, where there is a cold ocean current and a weak prevailing or synoptic wind regime. The mechanism is not fully understood, but Wexler (1946) and Defant (1951) offer fairly detailed descriptions. Sea breezes, created broadly during the period 0900 to 1900 by land surface heating and the consequent rising of hot low-density air over the land, often extend 30 to 50 km inland. They are usually much stronger and less variable than the counterpart night-time land breeze. Sea breezes as high as 35 m/s have been recorded but more typically in west coast tropical regions a mean value of 10 to 12 m/s (at 10 m height) can be expected. It is thought that land/sea systems are relatively shallow (less than 200 m depth) and they tend to divert around major topography; speed-up effects such as previously described over hills and ridges are not usually found.

The foregoing comments relate principally to the exploitation of either topographical features or special local thermodynamically induced effects to achieve high annual mean wind speeds.

Many potential users, because of the location of their power need, do not have the flexibility of siting that is necessary to take full advantage of topographical effects, and in such cases machines must of necessity be located in irregular or flat terrain where speed-up effects are transitory or non-existent. It is important therefore to consider other aspects of wind characteristics at prospective sites than simply annual mean wind speed.

Two of the most important features bearing upon wind turbine performance and integrity are the vertical wind gradient and the turbulence intensity.

Vertical wind gradient can be defined by a logarithmic or an exponential relationship (among others). The exponential form is generally used because of its simplicity, the wind speed V at height above ground Z (in relation to wind speed V_o at datum height Z_o) being given by

$$\frac{V}{V_o} = \left[\frac{Z}{Z_o}\right]^{1/\alpha} \quad (5.1)$$

The index $1/\alpha$ in Equation (5.1) is often called the vertical exponent; it is not only site-dependent but also varies with wind direction, wind speed, temperature, season and time of day. (Note that the α used here is different from the α used in Chapter 2 to denote angle of attack.) These various dependencies make α a highly statistical factor but one which is nevertheless useful because of its simplicity in use. Broadly, $1/\alpha$ can be expected to lie in the range 0.1 to 0.3, the lower values applying to exposed sites with relatively smooth surfaces in the prevailing direction, the higher values applying to more complex terrain with pronounced surface roughness, particularly urban areas. The importance of vertical gradient is twofold. For bigger horizontal-axis turbines (say 15 m diameter or more) a large vertical gradient can result in a substantial change in blade loading through one revolution. It is possible, even likely, that fatigue damage is a major contributor to failures of wind turbines in service and cyclic loads due to large vertical gradient could be a contributory factor to this damage.

The combined influence of α and turbine diameter D upon the ratio of mean wind speeds at the top (V_T) and bottom (V_B) of the turbine sweep is illustrated simply in Fig. 5.4. The range from practically unity (for a 1 m turbine with $\alpha = 0.1$) to nearly 1.4 (for a 15 m turbine with $\alpha = 0.3$) illustrates the importance of choosing low vertical gradient sites, especially for larger machines. Secondly, for all sizes of turbine the vertical gradient may influence the height of tower which is selected; a large vertical gradient will make a tall tower worth while, whereas a low vertical gradient will generally favour a shorter tower. The latter condition is generally found at the sites with highest mean wind speeds and this reinforces the preference towards low towers for structural integrity. Tower height could be introduced as a further sensitivity into Fig. 5.4 to explore this fully; since α is only a very simplified factor the virtue of pursuing such exploration is, however, dubious.

Equally important, but at present rather unquantified as regards its direct effect upon wind turbine integrity, is turbulence intensity I at a site. This is defined by

$$I_u = \frac{\sqrt{(u'^2)}}{\bar{u}} \quad (5.2)$$

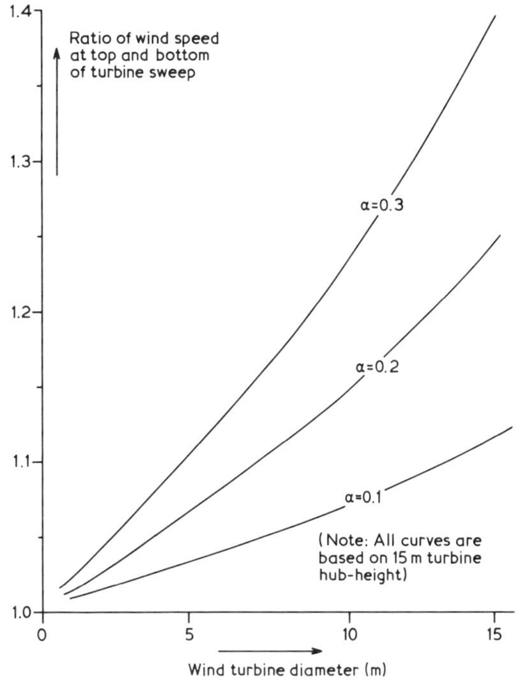

Fig. 5.4 Variation of wind speed across turbine disc, in relation to turbine diameter and exponent.

where I_u is the longitudinal turbulence intensity and $\bar{u}$ and u' are the mean and variant components of instantaneous longitudinal wind speed U, defined in Equation (5.3) by

$$U = \bar{u} + u' \quad (5.3)$$

Similar expressions are used for the lateral and vertical components of turbulence. Turbulence intensity is highest in irregular and rough terrain, lying approximately in the range 0.2 to 0.35 for such conditions. For smoother terrain it can be as low as 0.1. These figures relate to the longitudinal turbulence intensity; the lateral component is generally about 75 per cent of this value and the vertical component 50 per cent. Vertical wind gradient and turbulence intensity both depend upon surface roughness of the surrounding country, this usually being defined as a roughness length. Typical values of roughness length for different types of terrain are shown in Table 5.2, and the dependence of the vertical exponent $1/\alpha$ and turbulence intensity upon this range of roughnesses is also shown. A very good review of data concerning vertical gradient and turbulence intensity is given by Counihan (1975).

Superimposed on this general characterization of the wind at a site are specific conditions created by very local obstructions such as buildings and trees close to the site. Although attempts (notably by Wegley *et al.*, 1980) have been made to simplify the effects of such barriers and provide general guidance, it has to be said that the effects of such obstructions are very difficult to predict in detail, especially for natural winds from varying directions and with varying speeds. The safest way to avoid undesirable effects is to select sites as far from any notable obstructions as is practically possible, particularly in the direction of any prevailing wind.

Table 5.2 Typical surface roughness, vertical exponent and turbulence intensity for various types of terrain (Counihan, 1975).

Terrain type		Surface roughness (mm)	Vertical exponent, $1/\alpha$	Longitudinal turbulence intensity
Smooth	Ice	0.01	0.08	0.10
	Mud	0.4	↓	↓
	Snow	1.0	↓	↓
	Sea	2.0	0.12	0.12
Moderately rough	Short grass	1.0	0.13	0.13
	Grass and crops	30–70	0.143	↓
	Rural	200	0.16	0.20
Rough	Rural and woods	1000	0.20	
	Woods	↓	↓	0.20
	Suburbs	1500	0.23	
Very rough	Urban	1000–4000	0.25–0.40	0.30–0.40

One final point to be highlighted in this very brief review is the development of transition layers in the wind arising from changes in surface roughness upwind. Fig. 5.5 shows the irregularity that can occur in vertical wind gradient as a result of such roughness changes; such possibilities should be noted in prospective sites where serious roughness changes exist, and the influence the transition layer may have on wind speed at the turbine hub-height should be carefully considered. Fig. 5.6 gives simple guidance.

5.2 Wind measurement

Techniques for measuring the wind range from the very simple to the highly sophisticated, and the choice of method and instrument will depend upon the precise aspect of wind behaviour at a site which requires investigation.

For wind power projects or studies there are generally five facets of the wind regime within the overall likely height of the wind power installation at a prospective site which are important:

(1) The long-term characteristic, defined by the annual mean wind speed and by a wind speed–duration curve. This is important for the energy calculations described in Section 3.3.

(2) The extreme wind speed at the site. This is difficult to determine by short-term measurements and is usually defined in terms of a certain period of statistically likely recurrence, for example a 1 in 50 year return wind is quite commonly used.

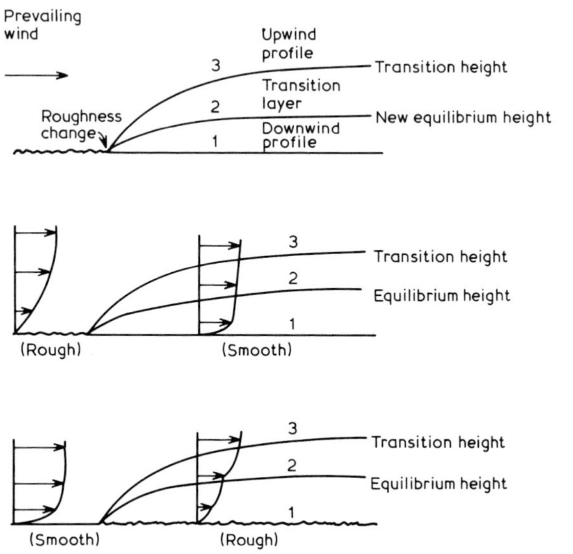

Fig. 5.5 Formation of a transition layer over changes in surface roughness.

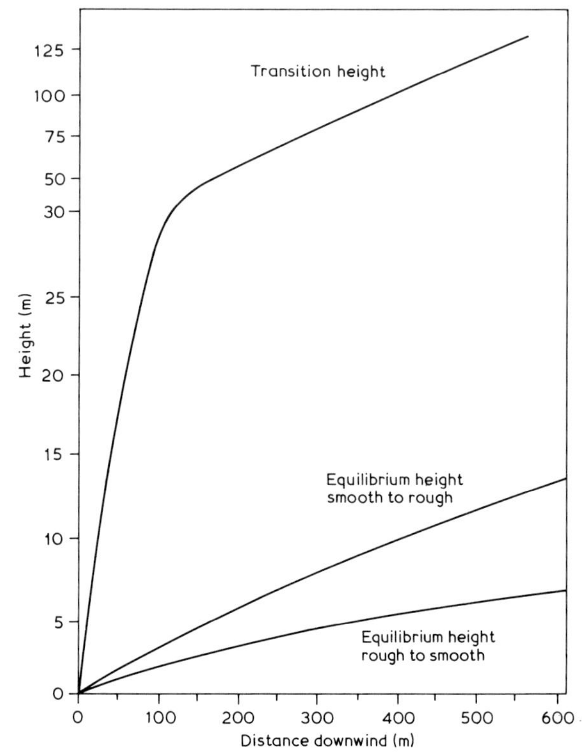

Fig. 5.6 Transition and equilibrium heights downwind of a change in surface roughness.

(3) Any cyclically repeatable or statistically likely patterns of wind speed and direction. This can be important when calculating energy storage requirements and when generally examining the match between wind power availability and user power demand. Diurnal variations are the most important but seasonal variations can also affect the selection and design of a wind power system.

(4) Turbulence. A large amount of turbulence at a site will increase the possibility of mechanical failure; unsteady wind conditions should also be avoided for smaller wind generators if rapid fluctuations in power output are undesirable.

(5) The vertical wind gradient. As discussed in the previous section this will have a bearing upon the preferred tower height, and in cases where the vertical gradient is pronounced and a large turbine is to be used, whether the site is acceptable because of excessive cyclic loading.

The following commentary on instruments and measurement is offered in relation to these five facets of wind behaviour. Discussion of sonic or laser Doppler anemometry, hot wire techniques and balloon mounted instruments, for instance, is omitted because the equipment is either more responsive than is necessary or may not be durable enough for field work, and it is usually too expensive for simple wind surveying purposes. These classes of instrument are also omitted from the reference section.

5.2.1 Observation

Measurements of natural wind first began several centuries ago. The earliest methods were based on observation and this is still used in some circumstances such as lighthouses and lightships. It is perhaps surprising that a skilled and experienced observer can, at an open site with reasonable scope for observation of tree movement or wave state, achieve accuracy of 10 per cent or better on wind speed and within a few degrees on wind direction.

The Beaufort scale was set out initially as guidance to the Royal Navy on observed wind conditions and has been extended for use on land. It gives twelve categories of wind speed which can readily be related, for instance, to wave conditions or tree motion, and a further five extremely high wind speed ranges. Table 5.3 shows the

Table 5.3 The Beaufort scale of wind force and specification (based on Simpson, 1906, and Golding, 1955).

Beaufort number (B)	Description of wind	Wind speed range at 10 m above ground (m/s)	Specification
0	Calm	less than 0.3	Smoke rises vertically
1	Light air	0.3–1.5	Wind direction shown by smoke drift but not by wind vanes
2	Slight breeze	1.6–3.3	Wind felt on face; leaves rustle; ordinary vane moved by wind
3	Gentle breeze	3.4–5.4	Leaves and twigs in constant motion; wind extends light flag
4	Moderate breeze	5.5–7.9	Raises dust and light paper; small branches are moved; white crests on waves
5	Fresh breeze	8.0–10.7	Small trees in leaf begin to sway; wavelets form on inland waters
6	Strong breeze	10.8–13.8	Large branches in motion; whistling heard in telegraph wires
7	Moderate gale	13.9–17.1	Whole trees in motion; inconvenience felt when walking against wind
8	Fresh gale	17.2–20.7	Breaks twigs off trees; generally impedes progress
9	Strong gale	20.8–24.4	Slight structural damage occurs (chimney pots and slates removed)
10	Whole gale	24.5–28.4	Seldom experienced inland; trees uprooted; considerable structural damage occurs
11	Storm	28.5–32.6	Very rarely experienced; accompanied by widespread damage
12	Hurricane	32.7–36.9	
13	—	37.0–41.4	
14	—	41.5–46.1	
15	—	46.2–50.9	
16	—	51.0–56.0	
17	—	56.1–61.2	

relationship of the various categories to ranges of wind speed. A relatively cheap method of estimating local wind speed is to make visual observations based on the guidance of the Beaufort scale over a few months, and to obtain measured data from the nearest recording meteorological station over the same period. A broad relationship between the two may then be assessed, and the long-term mean wind speed at the prospective site can be estimated, working from the long-term recorded mean wind speed from the meteorological station.

An extension of the observation technique for assessment of long-term mean wind speeds is the measurement of permanent vegetation distortion. There are four simple classifications of tree deformation as follows:

(1) *Brushing*, which is the most sensitive indicator of light winds. Branches and twigs are bent downwind as if combed or brushed; this tends to be demonstrated most clearly in deciduous trees when without leaves.

(2) *Flagging,* where branches stream downwind from the tree in a permanent deformation.

(3) *Throwing,* where not only branches but also the main trunk bend downwind.

(4) *Carpeting,* where the wind is so strong that twigs extending more than a few inches off the ground are killed and growth extends along the ground downwind in a form of carpet.

The 'calibration' of such deformities to estimate mean wind speed is probably affected by numerous factors, such as the tree type, temperature and humidity patterns, damage by other causes, soil conditions and the relative predominance of one single wind direction.

However, some progress has been made and Hewson, Wade and Baker (1978) have suggested mean wind speed ranges which might apply to the four classifications and sub-classifications of these. These ranges are reproduced in Table 5.4. It is to be hoped that in coming years the calibration of deformation can become increasingly sophisticated through further research, and it could then become a very valuable (and inexpensive) wind prospecting tool.

Table 5.4 Probable wind speed ranges for differing degrees of vegetation deformation.

Deformation classification	Probable annual mean wind speed range (m/s)*
Brushing and slight flagging	3.1–4.0
Slight flagging	4.0–4.9
Moderate flagging	4.9–5.8
Complete flagging (no upwind branches)	5.8–7.1
Partial throwing	6.7–8.0
Complete throwing	7.1–9.4

*Converted from rounded miles per hour figures; accuracy to 0.1 m/s is probably not realistic.

5.2.2 Instruments

Observation and the general guidelines for site selection set out in the previous sections are invaluable tools in making preliminary judgements on wind speeds. For some applications (wind pumping for instance) it may be that no further investigation is necessary; millions of wind pumps have been installed and successfully operated without wind measurements at the site. However, there are cases where it is important for the success of the wind power system to be more certain of the site wind characteristics. In these cases some form of instrument has to be called upon. The ensuing discussion of instrument types and their mode of use follows the five classes of wind data requirement listed at the beginning of Section 5.2.

For the long-term wind characteristic, the extreme wind speed and diurnal or seasonal wind patterns, wind sensors do not require fast response, but they must be durable because they may be left unattended for protracted periods. The most common form of speed sensor is the three-cup anemometer, an example of which is shown in Fig. 5.7. The cup anemometer was first developed by T.R. Robinson in 1846, his instrument being based upon four hemispherical cups. A great amount of theoretical and experimental investigation of this instrument has been carried out, and this has resulted (apart from the evolution towards three cups) in the introduction of beaded cup edges, the substitution of a conical for a hemispherical cup profile, and general redistribution of the arm and cup geometry. Most of the instruments listed in the reference section have adopted this principle and many of them incorporate the three conical cups discussed. Several different types of equipment embody a cup anemometer, each giving different forms of wind information.

The simplest and cheapest is the hand held indicating instrument. A number of companies manufacture these

Fig. 5.7 Three cup anemometer (by kind permission of Brookes and Gatehouse).

(Casella, Deuta, Munro and Simerl for example) and their use is the first logical extension of unaided observation techniques.

Mast mounted types can be divided into integrating, indicating and recording categories. Integrating anemometers consist of a rotor directly geared at the mast head to a mechanical counter, or a rotor driving a switch which breaks or makes once every certain number of revolutions (often 10), the open or close signal being used to drive an electromagnetic counter remote from the mast-head. The former type (represented in the reference list by the Casella W1200 and 1220 families and the Munro IM 119 group) are simple to erect, easy to use and reliable in service. The latter (often known as cup-contact instruments, exemplified by Vector Instruments A100R and Weather Measure W163/164 groups), is less simple to install because of the electrical wiring, but gives some flexibility in siting the counter. A variant on the cup-contact type is the Makino Ohyosokki AC750, in which a 'magne-contact' driven by the cup rotor operates a remote digital magnetic counter. In all integrating instruments the movement in counter reading (which represents the wind run in kilometres or miles past the instrument) over a certain period, if divided by that period, automatically gives the corresponding mean wind speed. The integrating instrument is therefore a cheap and convenient means of measuring long- or medium-term mean wind speed. It is less convenient (although not impossible with frequent readings) to obtain short-term means, such as hourly figures. This type of instrument is an ideal 'prospecting' device if the wind speed–duration curve shape is well known for the area and if there is no need to obtain detailed information on diurnal or seasonal variations. Indicating cup anemometer types incorporate a generator (either d.c. or permanent magnet a.c.) directly driven by the rotor, the signal being taken to some remote meter. Most companies in the reference list offer this type; for wind power work it is more convenient than the hand held indicating instrument, particularly at night or in bad weather, when a reading can be taken in protected conditions. Because there is no form of continuous information storage (as there is in integrating or recording instruments), this type is not particularly suitable for investigation of mean wind speeds.

The recording anemometer system has traditionally comprised an a.c. or d.c. generator driven directly by the three-cup rotor and linked electrically to a roll-chart recorder. The Makino Ohyosokki AF750, Simerl WS-77 and Société Précis Mécanique APM11/22 are examples among many of this type in the reference list. With the advent of light sensitive diodes and transistors and the development of digital circuit techniques the use of optical methods for sensing the rotor speed has become increasingly attractive in recent years. The general principle is to arrange for the shaft of the three-cup rotor to cyclically interrupt a light beam, the receiving light sensitive device then detecting the number and rate of revolutions. The signal is then capable of being processed digitally in a wide variety of ways. Apart from the increased flexibility of data processing, this technique permits slightly better response in the anemometer itself. There is, however, a need for more sophisticated power supplies or power translators. The Casella T16110/1, Climet 011-3 and MRI 1022 are examples of this 'photo-electric' or 'photo-chopper' system.

The recording instrument is especially useful where diurnal or seasonal wind patterns are to be investigated, where the wind speed–duration curve is required and in cases where there is general uncertainty about the temporal pattern of wind speeds at a site. Digital data processing clearly enables much greater power to analyse and manipulate the data, but if general investigatory work is necessary without a clear initial aim as to the data requirement and format, the analogue chart recording system may be preferable.

In the same 'general purpose' class of instrument as the rotating cup anemometer, but less common, are pressure tube and rotating vane devices.

The only examples in the reference list of a rotating vane anemometer (which has a rotor similar to the Savonius turbine described in Chapter 2) are those in the Simerl range. Vane rotors are simpler and more compact than their cup rotor counterparts, but they are less responsive.

The pressure tube anemometer has its origins in the work of W.H Dines. In the Dines anemometer a horizontal tube at the mast head is turned to face the wind by a vane; the horizontal tube leads into a vertical tube within the mast which in turn leads through further piping to the inside of a copper vessel floating in a cylindrical metal container, which is sealed from the air in the recording cabin. In addition, a little way down from the mast head the vertical pressure tube is surrounded by an outer concentric tube which is pierced on the outside by four rings of holes. This outer concentric tube is led separately to the air space above the copper vessel. As the wind speed rises, the pressure created inside the copper vessel is therefore assisted by the vacuum created above it in raising the level of the vessel, and this motion is transferred to the writing arm of a chart recorder. The Dines instrument has been extensively used for permanent meteorological station work and continues to be commercially available from Casella (model W1500) and Munro. However, the need for rigid mounting and the general construction of the instrument make it rather expensive and unsuitable for general wind survey work. Variants of the pressure tube principle are also available commercially from Dwyer of the USA. In the hand held Dwyer instrument the wind pressure supports a ball in a vertical glass tube, its position being read against a calibrated wind speed scale; this is one of the simplest commercially available anemometers.

Overlaying all the discussion on wind speed measurement, there is of course an associated need to measure wind direction. This is achieved in general purpose instruments with a direction vane. Direction signals are of little use if averaged over any period, so vanes are linked mainly to indicating or recording systems. In both cases some means of transmitting the signal is necessary, and both mechanical and electrical/electromagnetic methods are employed. An example of direct mechanical linkage by rod and crank is the Kahl 02 AM 300. Clearly the distance over which the signal can be transmitted by mechanical means is strictly limited and more commonly used are electromagnetic couplers of the magslip type, and potentiometers. In recent years conductive plastic potentiometers have been introduced and are now found in a number of commercial instruments (for example the Danjay AB 775D and the Société Précis Mécanique GPM23).

For investigations of turbulence, the response needs are greater than is available from the general purpose instruments. Also it is unlikely that turbulence studies at a site will persist for any substantial period, certainly nowhere approaching the length of service that is sometimes expected of a cup anemometer and direction vane for meteorological station use. In high response instruments, weight and durability therefore take lower priority. Before describing a few of the instruments available for this work it will be helpful to explain the way in which response is most conveniently described. The normal description of response is through a time constant, which is the period taken by the instrument to read $(1-1/e)$ of the full value of an applied step function. For anemometers this period is not an absolute number, but depends upon the level of wind speed imposed; it has therefore been found more convenient to introduce the concept of a distance constant. This is the length of wind flow which passes the anemometer before it reads $(1-1/e)$, or 63.2 per cent of the step change in wind speed. Gill and Hexter (1972) give a useful review, not only of response definition, but also accuracy, damping ratio, hysteresis, sensitivity and linearity definitions for wind measuring equipment. For general purpose cup anemometers the distance constant ranges between about 1.5 m and 20 m.

Fig. 5.8 R.M. Young Model 21004. Gill 'UVW' anemometer (by kind permission of R.M. Young).

Various types of instrument offer the potential to reduce this distance constant. First of all, the three-cup anemometer can be lightened and 'cleaned up' aerodynamically. The Casella T16110/1 and Young 12102 are examples of this. A more common solution to the need for higher response is the propeller type of anemometer represented in the reference list by the R.M. Young range. A three-propeller or UVW array is shown in Fig. 5.8. The distance constant of the devices in the Young range is 0.8 to 1.0 m, depending upon the size of propeller used. These instruments have been developed primarily to investigate the structure of light winds for instance in pollution dispersion studies, but it is quite possible that they will be equally useful in determining turbulence characteristics at wind power sites. Finally it is worth referring to the perforated sphere gust anemometer discussed by Golding (1955). This instrument comprises a lightweight perforated hollow sphere mounted on a stem capable of tilting in a single plane and having a strain gauge mounted on its hinge. It was developed further by ERA during the 1950s and 1960s and successfully and extensively used in field work. Although not fully commercially available details are still available from ERA.

Special attention is also needed to be given to direction vanes if they are to be used to investigate fine wind structure. The damping ratio h, given by

$$h = \frac{\text{Actual damping coefficient}}{\text{Critical damping coefficient}} \qquad (5.4)$$

is the important parameter. Vanes are underdamped (that is, their damping coefficient is less than the critical value which would just avoid overshoot), and so overshoot and oscillate for a short while about the new position. General purpose instruments have a damping ratio of about 0.2. Greater damping is required for diffusion and turbulence studies, and Gill and Hexter (1972) suggest that h should be in the region 0.35 to 0.7. The Munro IM 208 instrument is an example of increased damping; in place of the normal vane it incorporates twin fan-tails (similar to some of the larger wind generators discussed in Chapter 3 and detailed in the reference section).

The final aspect of wind behaviour listed as important at the beginning of Section 5.2 is the vertical wind gradient. The traditional solutions to determination of vertical gradient are tall towers with anemometers stationed at various heights, or various balloon techniques. Both of these are very expensive and normally inappropriate for a relatively small wind power installation. Recently a new instrument has been marketed which makes measurement up to 400 m relatively cheap.

The instrument, produced by Approach Fish Inc. and trademarked 'Tala', is based on a fabric calibrated kite. Operation is generally as for a normal kite, the wind speed being measured by the line tension, with due correction for catenary loads. An indicating version (using a spring gauge) and a tethered recording version (incorporating a strain gauge) are available. Preliminary checks against conventional instruments have shown reasonable accuracy and it seems quite possible that this type of instrument will find increasing use for wind power purposes. It may prove particularly effective in detecting the transition layers discussed at the end of Section 5.1.

This brief review of the requirements for wind data and the instrument types which are available to meet these needs has inevitably glossed over the considerable depth that can be found in the subject. In conclusion it should be emphasized that it is very important before undertaking a measurement programme to be quite clear what form of data is required; having decided this it is equally important (if instrumentation is to be used) to select the level and type of instruments appropriate to that need.

References

Counihan, J. (1975), Adiabatic atmospheric boundary layers; a review and analysis of data from the period 1890–1972, *Atmos. Environ.*, **9**, 871–905.

Defant, F. (1951), Local Winds, in *Compendium of Meteorology*, AMS, pp. 655–672.

Frenkiel, J. (1961), Wind flow over hills in relation to wind power utilisation, *Proceeding of the UN Conference on New Sources of Energy, Vol. 7: Wind Power,* UN, New York, pp. 85–111.

Gill, G.C. and Hextor, P.L. (1972), Some instrumentation definitions for use by meteorologists and engineers, *Second Symposium on Meteorological Observations,* San Diego, California.

Golding, E.W. (1955), *The Generation of Electricity by Wind Power,* E. & F.N. Spon, London.

Hewson, E.W., Wade, J.F. and Baker, R.W. (1978), *Vegetation as an Indicator of High Wind Velocity,* Oregon State University, Report RLO-2227-T24-78-2.

Simpson, G.C. (1906), *The Beaufort Scale of Wind Force,* Meteorological Office Official No. 180.

Tagg, J.R. (1957), *Wind Data Related to the Generation of Electricity by Wind Power,* Electrical Research Association, Leatherhead, UK, Report C/T 115.

Wegley, H.L., Ramsdell, J.V., Orgill, M.M. and Drake, R.L. (1980), *A Siting Handbook for Small Wind Energy Conversion Systems,* Battelle Pacific North West Laboratory Report PNL-2521 Rev. 1.

Wexler, R. (1946), Theory and observations of land and sea breezes, *Bull. Met. Soc.*, **27**, 272–287.

6 Commentary on the Reference Section

6.1 Classification

The reference section which follows is principally concerned with wind power equipment that is commercially available. The three main classifications are:

Wind generators
Wind pumps
Anemometers

The data have been obtained through a process of questionnaires which have been sent to all manufacturers known to be active in any of the above three categories. Detailed follow-up has been pursued directly with the manufacturer where necessary or appropriate, and all non-respondents to the questionnaire have been re-approached at least once. The entries are confined to information that has been provided directly by manufacturers or suppliers and recourse to other sources of data has been avoided in order to minimize erroneous or misleading entries. Regrettably this has resulted in some manufacturers which are known to be active not being represented by a full entry. In order to compensate, a list of such companies and their addresses accompanies the wind generator and wind pump sections. The following commentary provides some background on each of the three sections and guidance on their layout.

Lists of agents and intermediate suppliers have been omitted for the present. This has been done because of the relatively rapid pace of development in the field, and the correspondingly fluid situation that presently exists in dealerships, particularly for many types of wind generator. In these circumstances it seems most satisfactory for a prospective purchaser to commence enquiries with the original supplier and subsequently to be transferred to the current agent as appropriate.

6.2 Wind turbine generators

The present spectrum of manufacturers comprises some which have been established for forty or fifty years and others only in existence for a year or two. The present position is changing very rapidly, with new companies entering the field at quite a high rate, particularly in the USA and Denmark. It is quite likely that some of the newer companies which have either inadequate business footing or an unsound machine design will not survive, but which of the present list these will be is of course difficult to discern. Against this general background, a rather complicated decision has had to be made concerning the definition of 'commercial availability'. The marketing approach tends to differ between large and small companies, the former wishing to establish and consolidate their product through prolonged prototyping and trials before claiming 'in production', the latter often needing or wishing to sell even their first machine to a user as a commercial product.

Except where positively advised otherwise by the manufacturer, commercial availability for the purposes of this handbook constitutes more than one machine of any particular type built and commissioned.

It is recognized that valuable information would be lost from the handbook if the reference section were confined entirely to 'commercially available' machines, and a total of four categories for the wind generator section have been devised. These categories are as follows:

(1) Commercial machines (each identifiable design has a separate and detailed entry);
(2) One-off machines now built in the small/medium size range (compiled in a summary table);
(3) Small/medium size machines projected (compiled in a summary table);
(4) Large machines built or in process of being built (compiled in a summary table with selected photographs and drawings).

The final category has no direct bearing on the aims or likely readership of the book, but it is hoped that its inclusion will set some perspective on the overall wind power R & D position. The second category includes some machines which are probably commercially available, but on which insufficient data were found.

At an early stage of the handbook preparation separate entries for component suppliers were envisaged. This would have included such items as blades, towers and generators. In accord with the philosophy finally adopted and set out in the introduction (that of presenting the prospective user with the available range of complete equipment only) it was decided in the event to omit component suppliers.

The data sheets for wind turbine generators are separated principally into operational and performance details, main physical details including dimensions, weights and layout, and such production data as are available. The production information dates from when it was collected, and is accurate as regards machines produced for a range of dates from late 1979 to mid 1982.

Photographs, drawings and power–wind speed characteristics accompany each data sheet wherever these have been made available by the manufacturer. The cautionary notes made in Chapter 3 concerning the accuracy, resolution and repeatability of these curves should be borne in mind when considering these data. It should be noted that rated power denotes generally the electrical and mechanical duty that can be continuously sustained. The peak power delivery of a wind generator will under some conditions exceed this rated power for short periods.

The survival wind speed figure, where quoted, should be treated with caution. Some manufacturers indicate that the figure holds only for steady condition, others are not so specific. Rarely has there been adequate testing to define this figure accurately (assuming even that it is possible to do this without numerous conditions and riders). No comment is offered on the strengths or weaknesses of machines listed or on any documented (or other) operational experience.

The upper size and rating limit covered by the reference list has been set automatically by the definition of 'commercial availability', particularly if machines developed under government contract (as are most of the present large machines) are omitted. It is possible that any future revision of the list will either need upper size limits imposed, or have to include very large machines as these enter the phase of normal commercial dealings between manufacturer and electric utility. For the moment such large machines can, however, be confined to the final subsection of the reference list as it is presently set out.

6.3 Wind pumps

The wind pump data are set out in the form of the traditional pumping capacity table for each range of wind pumps. Where supplied by the manufacturer, photographs and/or diagrams of a typical machine from each range accompany these tables.

Most manufacturers supply pumping data in imperial units and in the interests of rationalizing on SI these have been converted to metres (lift) and cubic metres per hour (capacity). The appropriate conversion factors are listed in the Appendix. It has not always been clear whether manufacturers' data have referred to US gallons or imperial gallons; certain assumptions have been necessary in making the conversions. Most manufacturers still work with imperial units for turbine diameter (6 ft, 10 ft, 12 ft, etc.) and the conversion of these has resulted in unrounded metric dimensions. It is noted that the more modern ranges of European wind pumps use SI units as a base; no doubt this practice will spread as new designs and ranges emerge.

Most manufacturers qualify their pumping tables with a comment that capacities are approximate, and this is reproduced in the tables here. The comment made in Chapter 4 concerning mean or instantaneous wind speed datum is reinforced; it is not clear from most manufacturers' literature whether the pumping capacities are average figures that can be expected in the nominated mean wind speed, or whether it is the instantaneous capacity at that instantaneous wind speed. There could be a significant difference between the two.

6.4 Anemometers

The main information provided for each instrument model is the range, accuracy, response, type of transducer and form of output. It has clearly been inappropriate to convert data on range etc. from the manufacturers' units to SI, and so original units appear. This has resulted in a mixture of such units as metres per second, kilometres per hour, miles per hour and knots in the table. It is hoped that this mixture does not impede the ease of reference to the data.

Only those types of instruments thought to be useful for wind power purposes have been included. Hot wire, laser Doppler and sonic anemometers are omitted, and so also are the instruments designed especially for low-speed, axial-flow ventilation monitoring. Instruments primarily intended for glasshouse control and crane mounting are included because of their broad applicability to wind power work.

A number of companies specialize and deal in a broad range of scientific instruments, and these companies often factor anemometers made by other organizations. Double entries for one instrument have been avoided where possible, and where a choice exists, the instrument has been entered under the original manufacturer's name.

Some specialist electronics companies are now manufacturing data processing equipment to match a particular type of proprietary anemometer. The equipment, however, is too diverse in nature to tabulate or describe briefly.

7 Reference section

7.1 Wind turbine generators

7.1.1 *Commercial wind turbine generators*

Aero Lectric C9 Wind Wizard

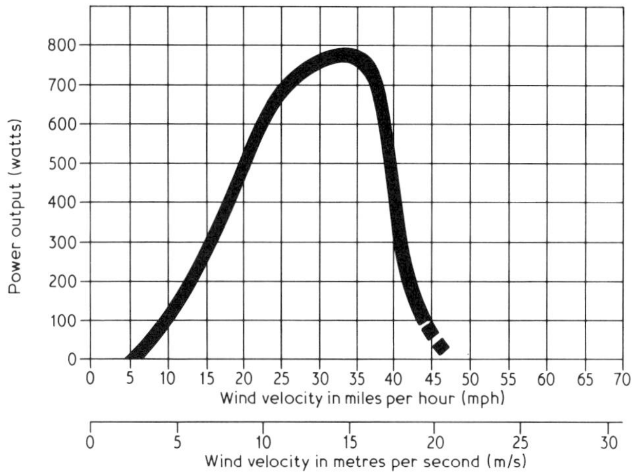

PERFORMANCE AND OPERATIONAL DATA

Rated electrical power	600 watt
Cut-in wind speed	4.0 m/s
Rated wind speed	11.6 m/s
Shut-down wind speed (adjustable)	11.2–17.8 m/s
Survival wind speed	40.1 m/s
Shaft speed at rated output	377 rev/min
Maximum shaft speed	720 rev/min

PHYSICAL DATA

Turbine

Type: Horizontal axis, upwind with tail vane

Dimensions

diameter 2.75 m

Blades

Construction: Wood with urethane finish or fibreglass and composite material

Number: 3

Mechanical transmission: Speed increasing V-belt drive

System weight: 22.6 kg (C9C model), 36.3 kg (C9D model)

MISCELLANEOUS

One year warranty on parts and workmanship.
Owners manual available.
Semi-annual inspection.

VARIANTS

C9C model is portable, with quick-mount detachable blade system.
C9D model designated as 'permanent'.

Aero Lectric Company,
13517 Winters Avenue, Cresaptown,
Maryland 21502
USA
Telephone: (301) 724 9165

Aero Power SL 1000

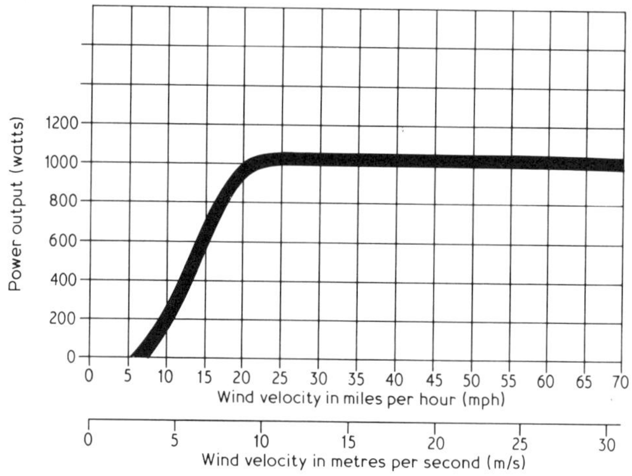

PERFORMANCE AND OPERATIONAL DATA

Rated electrical power	1000 watt
Rated output voltage	12/24/48 volt d.c.
Cut-in wind speed	2.7 m/s
Rated wind speed	10.0 m/s
Survival wind speed	44.7 m/s
Maximum shaft speed	475 rev/min

PHYSICAL DATA

Turbine
Type: Horizontal axis, upwind with tail vane
Dimensions
diameter 3.05 m
Blades
Construction: Sitka spruce, acrylic enamel painted with stainless-steel tape covering on leading edge
Number: 3
Control: Centrifugally operated blade pitch control
Generator: 3 phase a.c. generator
Mechanical transmission: 3:1 speed increasing high torque drive (HTD) system
System weight
gross 71.2 kg

PRODUCTION

First production: 1977
Production quantity to date: About 150 (October 1979)

MISCELLANEOUS

Unarco–Rohn towers are offered.
Aero Power sinewave inverter is offered.
Defects in workmanship and/or materials are covered under warranty for one year from purchase.
Owners manual including assembly, wiring and maintenance is available.
Semi-annual maintenance includes greasing of hub, checking blades for nicks and scratches, inspection of complete systems.
Shut-down achieved by turning turbine into plane of tail vane and wind. Manually operated, but automatic system available as extra.

REPORTS

Rocky Flats Performance Summary Sheet: Aero Power SL 1000, December 1980.

Aero Power Systems Inc.,
2398 Fourth Street, Berkeley,
California 94710
USA
Telephone: (415) 848 2710

Aero Power SL 1500

PERFORMANCE AND OPERATIONAL DATA

Rated electrical power	1500 watt
Rated output voltage	14/28 volt d.c.
Cut-in wind speed	3.3 m/s
Rated wind speed	10.0 m/s
Shut-down wind speed	44.7 m/s
Survival wind speed	44.7 m/s
Maximum shaft speed	375 rev/min

PHYSICAL DATA

Turbine

Type: Horizontal axis, upwind with tail vane
Dimensions
diameter 3.66 m

Blades

Construction: Sitka spruce, acrylic enamel painted with stainless steel tape covering on leading edge
Number: 3

Control: Centrifugally operated blade pitch control
Generator: 3 phase a.c. generator
Mechanical transmission: 3:1 speed increasing high torque drive (HTBD) system

System weight

gross 81.0 kg

PRODUCTION

First production: 1978 (HTBD replaced gear drive in early 1979)
Production quantity to date: About 100 (October 1979)

MISCELLANEOUS

Unarco–Rohn towers are offered.
Aero Power sinewave inverter is offered.
Synchronous inverter with air core reactor is available to operate the SL 1500 on the network.
Defects in workmanship and/or materials are covered under warranty for one year from purchase.
Owners manual including assembly, wiring and maintenance is available.
Semi-annual maintenance includes greasing of hub, checking blades for nicks and scratches, inspection of complete system.
Shut-down achieved by turning turbine into plane of tail vane and wind. Manually operated, but automatic system available as extra.

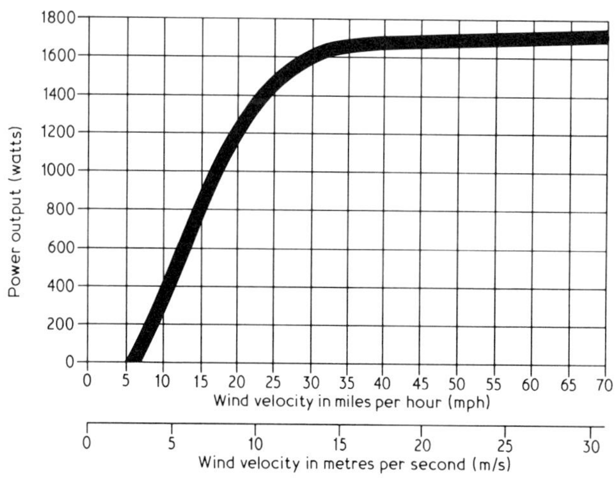

Aero Power Systems Inc.,
2398 Fourth Street, Berkeley,
California 94710
USA
Telephone: (415) 848 2710

Aerowatt 60 FP7G

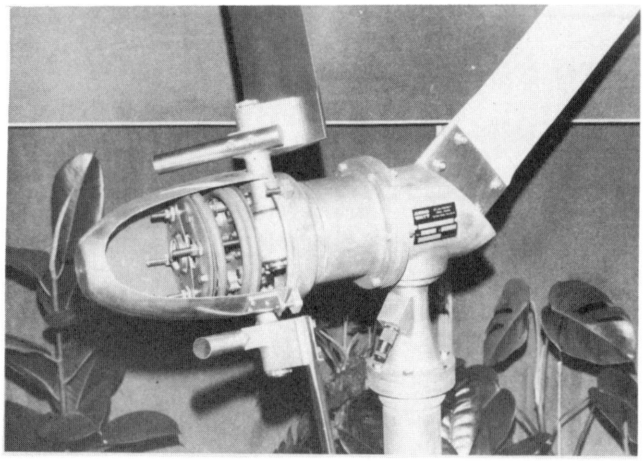

PERFORMANCE AND OPERATIONAL DATA

Rated electrical power	60 watt
Rated output voltage	6/12/24/36/48 volt d.c.
Cut-in wind speed	2.5 m/s
Rated wind speed	7 m/s
Shut-down wind speed	none
Survival wind speed	90 m/s
Shaft speed at rated output	900 rev/min
Maximum shaft speed	950 rev/min

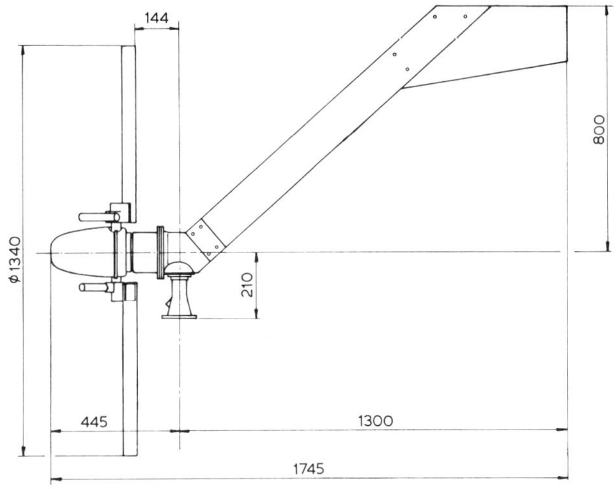

All dimensions in mm

PHYSICAL DATA

Turbine

Type: Horizontal axis, upwind with tail vane
Dimensions
diameter 1.34 m

Blades

Construction: Beechwood, polyurethane coated
Number: 2

Control: Centrifugally operated blade pitch control

Generator: 2-phase permanent magnet a.c. generator (rectified)

Mechanical transmission: Direct coupling to turbine

System weight

net 22 kg
gross 58 kg

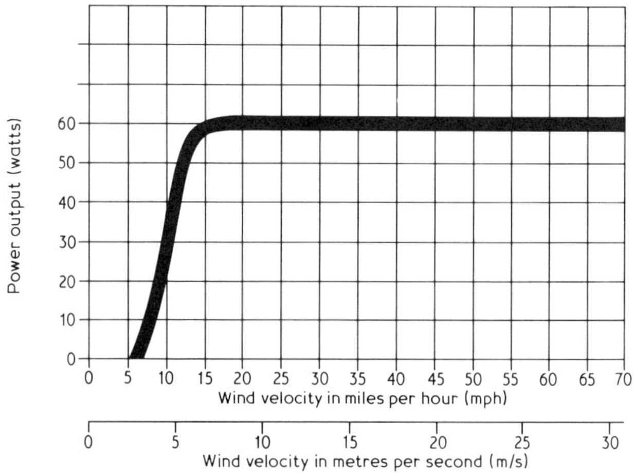

PRODUCTION

First production: Early 1979
Production quantity to date: In the order of tens

MISCELLANEOUS

Annual maintenance, comprising overall inspection, checking nuts, screws, bolts and guys for correct tightness.

Five year warranty on wind generator, one year on electronics. Parts and labour covered if returned to AW factory, otherwise parts only.

Handbook supplied with equipment.

Aerowatt
37, rue Chanzy,
75011 Paris,
France
Telephone: (371) 35 78
(371) 36 21
Telex: SYSNA 680 160F

Aerowatt 150 FP7G

PERFORMANCE AND OPERATIONAL DATA

Rated electrical power	150 watt
Rated output voltage	12/24/36/48 volt d.c.
Cut-in wind speed	2 m/s
Rated wind speed	7 m/s
Shut-down wind speed	none
Survival wind speed	60 m/s
Shaft speed at rated output	550 rev/min
Maximum shaft speed	575 rev/min

PHYSICAL DATA

Turbine
Type: Horizontal axis, upwind with tail vane
Dimensions
diameter 2.0 m
Blades
Construction: Extruded aluminium alloy
Number: 2
Control: Centrifugally operated blade pitch control
Generator: Permanent magnet a.c. generator (rectified)
Mechanical transmission: Direct coupling to turbine
System weight
net 80 kg
gross 183 kg

PRODUCTION

First production: 1973/4

MISCELLANEOUS

Solid-state voltage regulation systems type AW 150 RP 12, 24, 36 or 48 are offered.
Tiltable tower type SH 20A/120/Adx/Zn is offered, or free-standing type.
Teflon blade coating is available for sites where icing is anticipated.
Annual maintenance, comprising overall inspection, checking nuts, screws, bolts and guys for correct tightness.
Five year warranty on wind generator, one year on electronics. Parts and labour covered if returned to AW factory, otherwise parts only.
Handbook supplied with equipment.

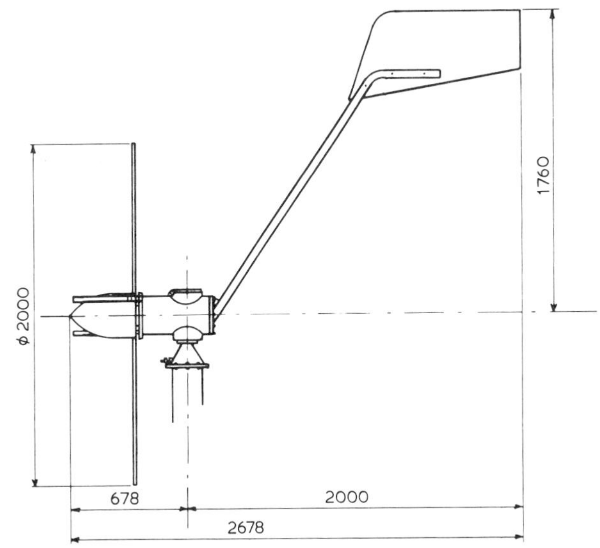

All dimensions in mm

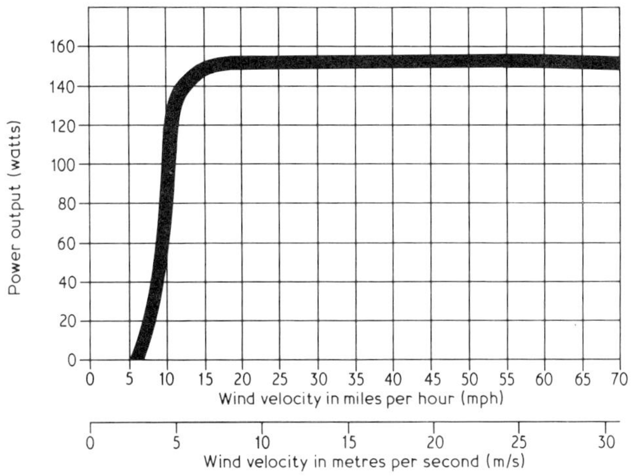

Aerowatt
37, rue Chanzy,
75011 Paris,
France
Telephone: (371) 35 78
(371) 36 21
Telex: SYSNA 680 160F

Aerowatt 100 FP5G

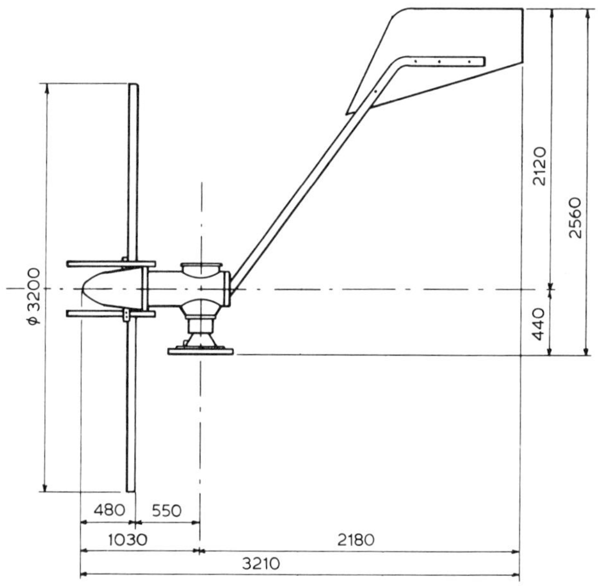

All dimensions in mm

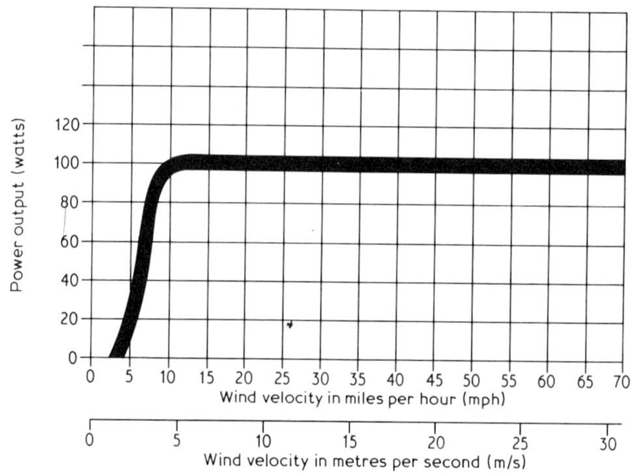

PERFORMANCE AND OPERATIONAL DATA

Rated electrical power — 100 watt
Rated output voltage — 12/24/36/48 volt d.c.
Cut-in wind speed — 1.5 m/s
Rated wind speed — 5 m/s
Shut-down wind speed — none
Survival wind speed — 60 m/s
Shaft speed at rated output — 380 rev/min
Maximum shaft speed — 395 rev/min

PHYSICAL DATA

Turbine
Type: Horizontal axis, upwind with tail vane
Dimensions
diameter 3.2 m
Blades
Construction: Extruded aluminium alloy
Number: 2
Control: Centrifugally operated blade pitch control
Generator: Permanent magnet a.c. generator (rectified)
Mechanical transmission: Direct coupling to turbine
System weight
net 176 kg
gross 345 kg

MISCELLANEOUS

Solid-state voltage regulation systems type AW 100 RP 12, 24, 36 or 48 are offered.
Tiltable tower type SH 32B/120/Adx/ZQ is offered, or free-standing type
Teflon blade coating is available for sites where icing is anticipated.
Annual maintenance, comprising overall inspection, checking nuts, screws, bolts and guys for correct tightness.
Five year warranty on wind generator, one year on electronics. Parts and labour covered if returned to AW factory, otherwise parts only.
Handbook supplied with equipment.

Aerowatt
37, rue Chanzy,
75011 Paris,
France
Telephone: (371) 35 78
(371) 36 21
Telex: SYSNA 680 160F

Aerowatt 300 FP7G

PERFORMANCE AND OPERATIONAL DATA

Rated electrical power	300 watt
Rated output voltage	12/24/36/48 volt d.c.
Cut-in wind speed	2.5 m/s
Rated wind speed	7 m/s
Shut-down wind speed	none
Survival wind speed	60 m/s
Shaft speed at rated output	420 rev/min
Maximum shaft speed	435 rev/min

PHYSICAL DATA

Turbine
Type: Horizontal axis, upwind with tail vane
Dimensions
diameter 3.2 m

Blades
Construction: Extruded aluminium alloy
Number: 2

Control: Centrifugally operated blade pitch control

Generator: Permanent magnet a.c. generator (rectified)

Mechanical transmission: Direct coupling to turbine

System weight
net 176 kg
gross 345 kg

PRODUCTION

First production: 1973/4

MISCELLANEOUS

Solid-state voltage regulation systems type AW 300 RP 12, 24, 36 or 48 are offered.
Tiltable tower type SH 32B/120/Adx/ZN is offered, or free-standing type.
Teflon blade coating is available for sites where icing is anticipated.
Annual maintenance, comprising overall inspection, checking nuts, screws, bolts and guys for correct tightness.
Five year warranty on wind generator, one year on electronics. Parts and labour covered if returned to AW factory, otherwise parts only.
Handbook supplied with equipment.
This machine is obsolescent. It will be replaced by a new model 1000 FP9, rated 1000 watt (a.c.) at 9 m/s (dimensionally similar).

Aerowatt
37, rue Chanzy,
75011 Paris,
France
Telephone: (371) 35 78
(371) 36 21
Telex: SYSNA 680 160F

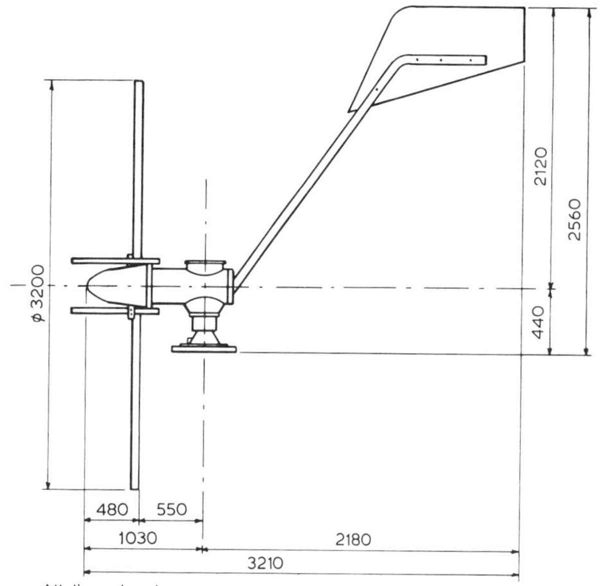

All dimensions in mm

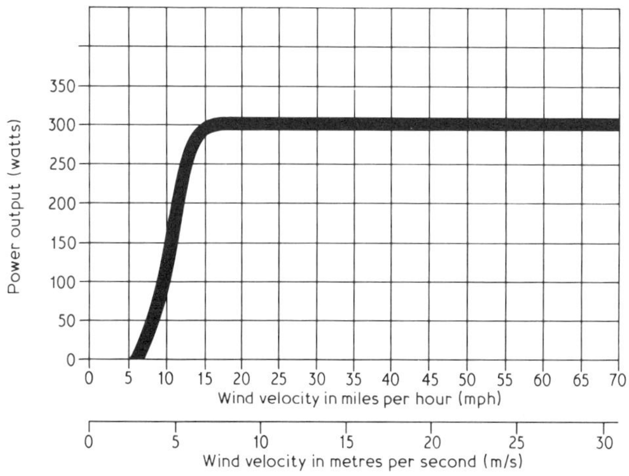

Aerowatt 1200 FP10G

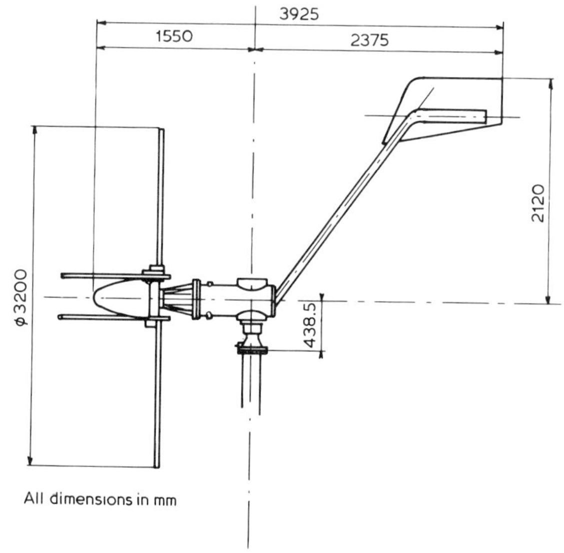

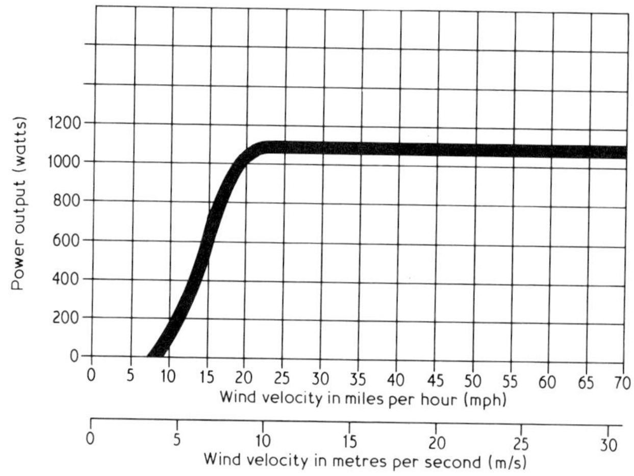

PERFORMANCE AND OPERATIONAL DATA

Rated electrical power	1080 watt
Rated output voltage	24/36/48/120 volt d.c.
Cut-in wind speed	3.5 m/s
Rated wind speed	10 m/s
Shut-down wind speed	none
Survival wind speed	60 m/s
Shaft speed at rated output	420 rev/min
Maximum shaft speed	435 rev/min

PHYSICAL DATA

Turbine
Type: Horizontal axis, upwind with tail vane
Dimensions
diameter 3.2 m
Blades
Construction: Extruded aluminium alloy
Number: 2
Control: Centrifugally operated blade pitch control
Generator: Permanent magnet a.c. generator (rectified)
Mechanical transmission: Coaxial speed increasing gearbox
System weight
net 220 kg
gross 390 kg

MISCELLANEOUS

Unrectified generator output available as an option, 1200 VA rating, 220–380 volt, 50 Hz.
Solid-state voltage regulation systems type AW 1100 RP 12, 24, 36 or 48 or AW 1100 RS 120 are offered.
Tiltable tower type SH 32B/120/Adx/ZN is offered, or free-standing type.
Polyethylene blade coating is available for sites where icing is anticipated.
Annual maintenance, comprising overall inspection, change of gearbox oil, checking nuts, screws, bolts and guys for correct tightness.
Five year warranty on wind generator, one year on electronics. Parts and labour covered if returned to AW factory, otherwise parts only.
Handbook supplied with equipment.
This machine is obsolescent. It will be replaced by a new model 1000 FP9, rated 1000 watt (a.c.) at 9 m/s (dimensionally similar).

Aerowatt
37, rue Chanzy,
75011 Paris,
France
Telephone: (371) 35 78
(371) 36 21
Telex: SYSNA 680 160F

Aerowatt 1100 FP7G

PERFORMANCE AND OPERATIONAL DATA

Rated electrical power	960 watt
Rated output voltage	24/36/48/120 volt d.c.
Cut-in wind speed	2.5 m/s
Rated wind speed	7 m/s
Shut-down wind speed	none
Survival wind speed	60 m/s
Shaft speed at rated output	178 rev/min
Maximum shaft speed	186 rev/min

PHYSICAL DATA

Turbine
Type: Horizontal axis, upwind with tail vane
Dimensions
diameter 5.0 m
Blades
Construction: Extruded aluminium alloy
Number: 2
Control: Centrifugally operated blade pitch control
Generator: Permanent magnet a.c. generator (rectified)
Mechanical transmission: Coaxial speed increasing gearbox
System weight
net 350 kg
gross 515 kg

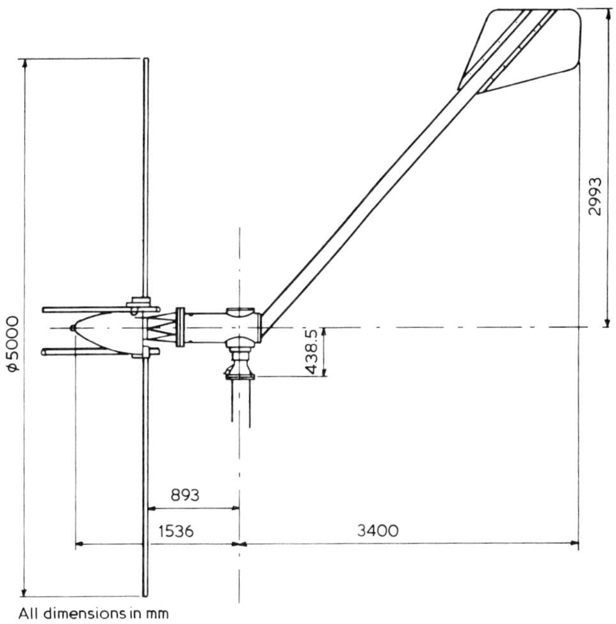

PRODUCTION

First production: 1973/4

MISCELLANEOUS

Unrectified generator output available as an option, 1250 VA rating, 220–380 volt, 50 Hz.
Solid-state voltage regulation systems type AW 1100 RP 24, 36 or 48 or AW 1100 RS 120 are offered.
Tiltable tower type SH 50B/150/Adx/ZN is offered, or free-standing type.
Polyethylene blade coating is available for sites where icing is anticipated.
Annual maintenance, comprising overall inspection, change of gearbox oil, checking nuts, screws, bolts and guys for correct tightness.
Five year warranty on wind generator, one year on electronics. Parts and labour covered if returned to AW factory, otherwise parts only.
Handbook supplied with equipment.

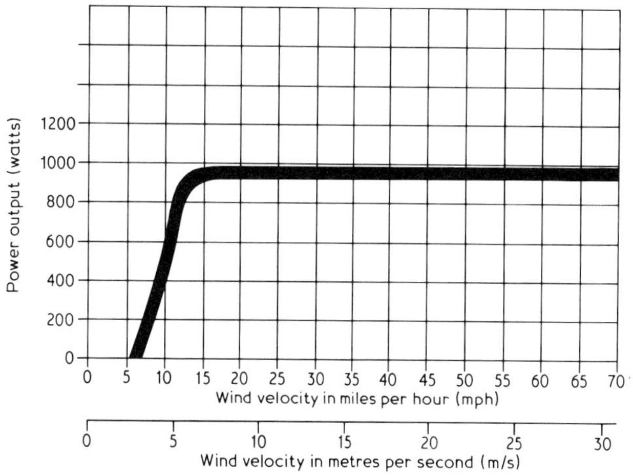

Aerowatt
37, Rue Chanzy,
75011 Paris,
France
Telephone: (371) 35 78
(371) 36 21
Telex: SYSNA 680 160F

Aerowatt 1100 FP5G

PERFORMANCE AND OPERATIONAL DATA

Rated electrical power	960 watt
Rated output voltage	24/36/48/120 volt d.c.
Cut-in wind speed	1.5 m/s
Rated wind speed	5 m/s
Shut-down wind speed	none
Survival wind speed	60 m/s
Shaft speed at rated output	130 rev/min
Maximum shaft speed	135 rev/min

PHYSICAL DATA

Turbine
Type: Horizontal axis, upwind with tail vane
Dimensions
diameter 9.2 m

Blades
Construction: Extruded aluminium alloy
Number: 2

Control: Centrifugally operated blade pitch control

Generator: Permanent magnet a.c. generator (rectified)

Mechanical transmission: Coaxial speed increasing gearbox

System weight
net 945 kg
gross 1530 kg

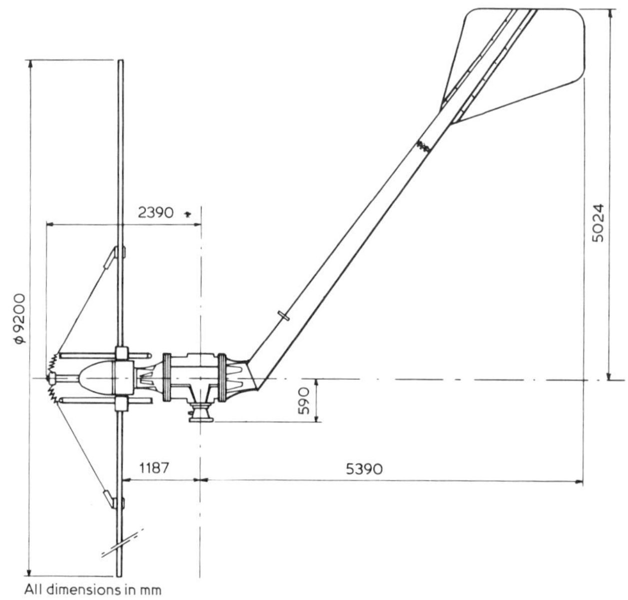

MISCELLANEOUS

Unrectified generator output available as an option, 1250 VA rating, 220–380 volt, 50 Hz.
Solid-state voltage regulation systems type AW 1100 RP 24, 36 or 48 or AW 1100 RS 120 are offered.
Tiltable tower type SH 92C/150/Adx/ZQ is offered, or free-standing type.
Polyethylene blade coating is available for sites where icing is anticipated.
Annual maintenance, comprising overall inspection, change of gearbox oil, checking nuts, screws, bolts and guys for correct tightness.
Five year warranty on wind generator, one year on electronics. Parts and labour covered if returned to AW factory, otherwise parts only.
Handbook supplied with equipment.

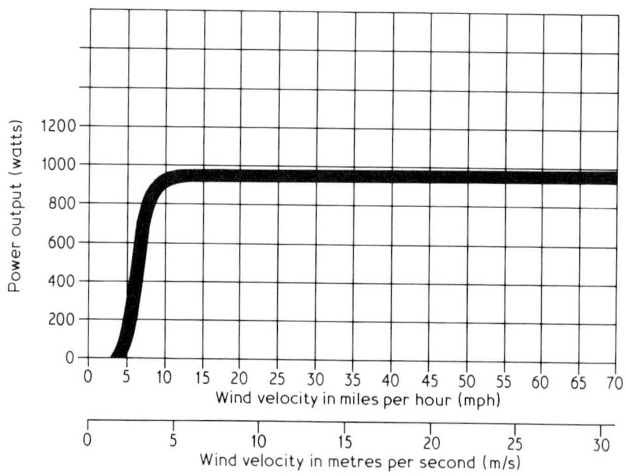

Aerowatt
37, Rue Chanzy,
75011 Paris,
France
Telephone: (371) 35 78
(371) 36 21
Telex: SYSNA 680 160F

Aerowatt 4100 FP7G

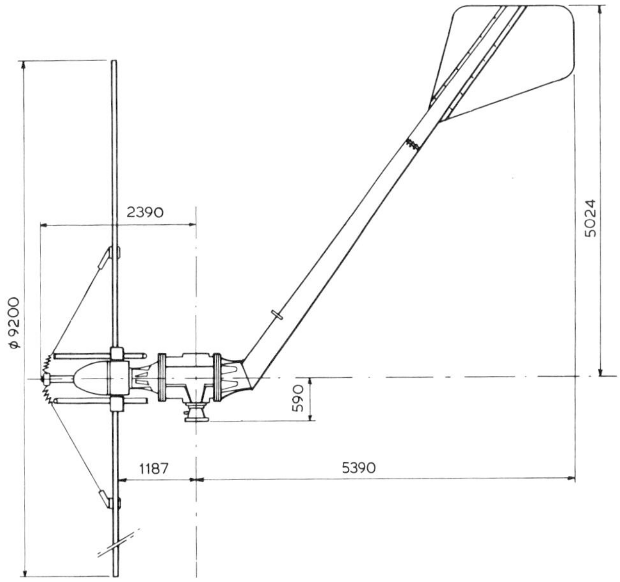

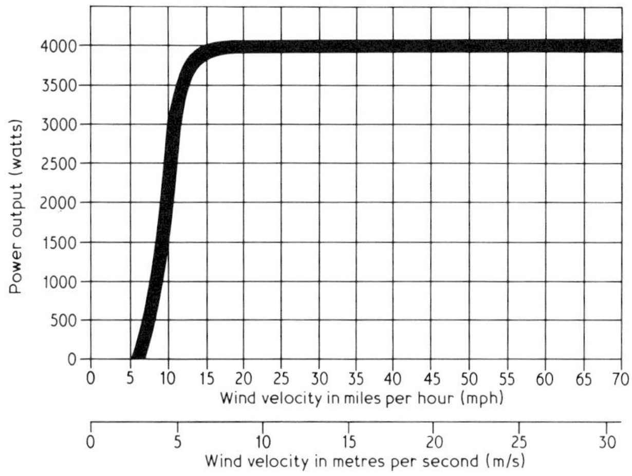

PERFORMANCE AND OPERATIONAL DATA

Rated electrical power	4000 watt
Rated output voltage	48/120/240 volt d.c.
Cut-in wind speed	2.5 m/s
Rated wind speed	7 m/s
Shut-down wind speed	none
Survival wind speed	60 m/s
Shaft speed at rated output	152 rev/min
Maximum shaft speed	160 rev/min

PHYSICAL DATA

Turbine

Type: Horizontal axis, upwind with tail vane
Dimensions
diameter 9.2 m

Blades

Construction: Extruded aluminium alloy
Number: 2

Control: Centrifugally operated blade pitch control

Generator: Permanent magnet a.c. generator (rectified)

Mechanical transmission: Coaxial speed increasing gearbox

System weight
net 971 kg
gross 1550 kg

PRODUCTION

First production: 1973/4

MISCELLANEOUS

Unrectified generator output available as an option, 4500 VA rating, 220–380 volt, 50 Hz.
Solid-state voltage regulation systems type AW 4100 RS 48, 120 or 240 are offered.
Tiltable tower type SH 92C/150/Adx/ZN is offered, or free-standing type.
Polyethylene blade coating is available for sites where icing is anticipated.
Annual maintenance, comprising overall inspection, change of gearbox oil, checking nuts, screws, bolts and guys for correct tightness.
Five year warranty on wind generator, one year on electronics. Parts and labour covered if returned to AW factory, otherwise parts only.
Handbook supplied with equipment.

Aerowatt
37, Rue Chanzy,
75011 Paris,
France
Telephone: (371) 35 78
(371) 36 21
Telex: SYSNA 680 160F

Alcoa Alvawt 835524

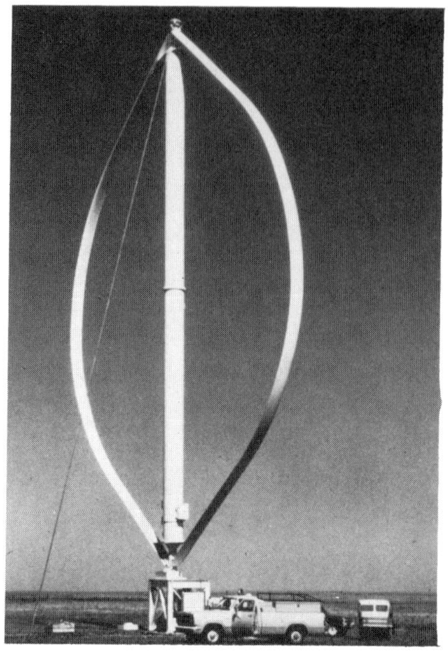

PERFORMANCE AND OPERATIONAL DATA

Rated electrical power	100 000 watt
Rated output voltage	480 volt a.c.
Cut-in wind speed (site adjustable)	5–7 m/s
Rated wind speed	13–14 m/s
Shut-down wind speed (site adjustable)	14–27 m/s
Survival wind speed	45–70 m/s
Shaft speed at rated output	48 rev/min
Maximum shaft speed	48 rev/min

PHYSICAL DATA

Turbine
Type: Vertical axis 'troposkien'
Dimensions
height 25 m
diameter 17 m

Blades
Construction: Extruded aluminium
Number: 2

Control: Fixed pitch blades
Generator: Induction generator
Mechanical transmission: Speed increasing transmission
System weight: 12 000 kg

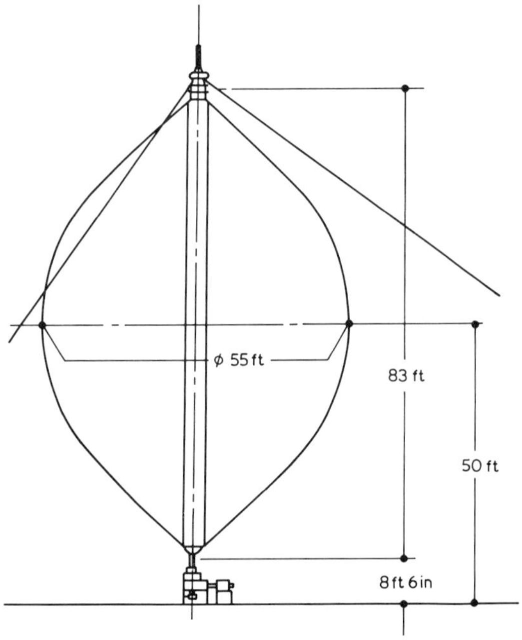

PRODUCTION

First production: 1980
Production quantity to date: 4 (May 1981)

MISCELLANEOUS

Maintenance requirements quoted as 'site specific'.
Guarantees or warranty quoted as 'project specific'.
Design variants quoted as 'project specific'.

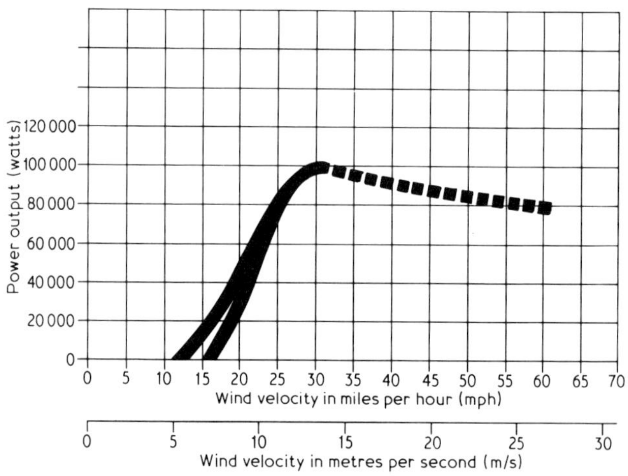

ALCOA Allied Products,
ALCOA Technical Center,
ALCOA Center,
PA 15069
USA
Telephone: (412) 339 6651

Alcoa Alvawt 1238229

PERFORMANCE AND OPERATIONAL DATA

Rated electrical power	500 000 watt
Rated output voltage	480 volt a.c.
Cut-in wind speed (site adjustable)	5–7 m/s
Rated wind speed	15–17 m/s
Shut-down wind speed (site adjustable)	14–27 m/s
Survival wind speed	45–70 m/s
Shaft speed at rated output	41 rev/min
Maximum shaft speed	41 rev/min

PHYSICAL DATA

Turbine
Type: Vertical axis 'troposkien' type
Dimensions
height 37 m
diameter 25 m
Blades
Construction: Extruded aluminium
Number: 3
Control: Fixed pitch blades
Generator: Induction generator
Mechanical transmission: Speed increasing gearbox

PRODUCTION

First production: 1980
Production quantity to date: 3 (May 1981)

MISCELLANEOUS

Maintenance requirements quoted as 'site specific'.
Guarantees or warranty quoted as 'project specific'.
Design variants quoted as 'project specific'.

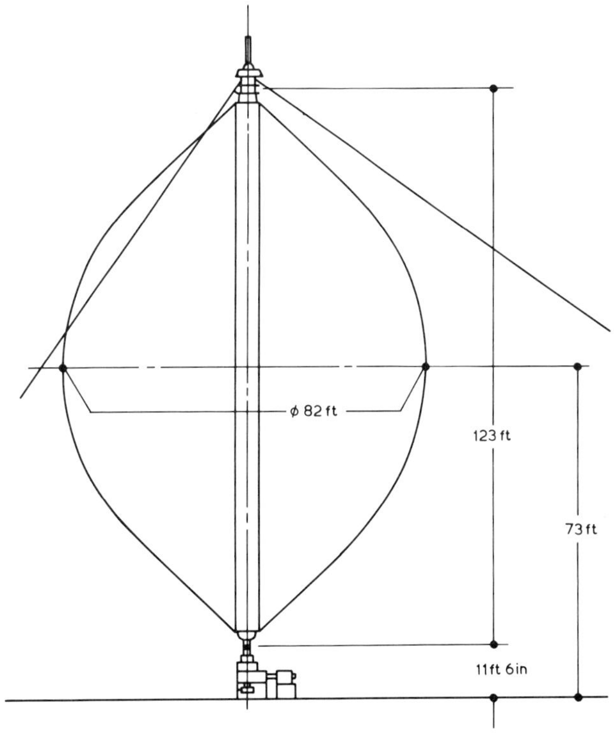

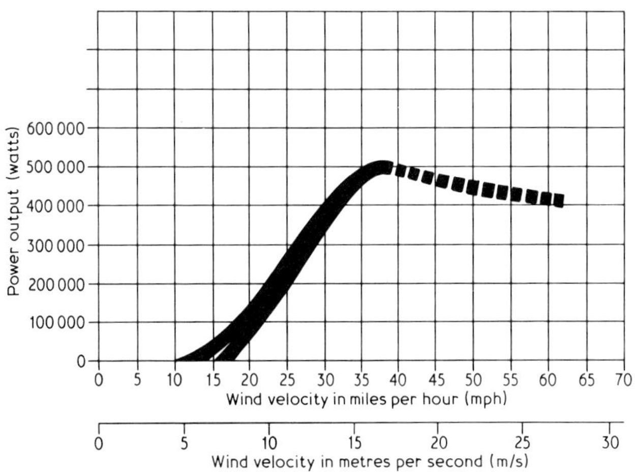

ALCOA Allied Products,
ALCOA Technical Center,
ALCOA Center,
PA 15069
USA
Telephone: (412) 339 6651

Alternative Energy Ltd 500 watt

PERFORMANCE AND OPERATIONAL DATA

Rated electrical power	500 watt
Rated output voltage	24 volt
Cut-in wind speed	3.1 m/s
Rated wind speed	9.4 m/s
Shut-down wind speed	22.3–24.5 m/s
Survival wind speed (tested to 29.4 m/s)	49.1 m/s
Shaft speed at rated output	350 rev/min
Maximum shaft speed	450 rev/min

PHYSICAL DATA

Turbine
Type: Horizontal axis, upwind with tail vane
Dimensions
diameter 2.3 m
Blades
Construction: Laminated spruce
Number: 3
Control: Fixed pitch turbine
Mechanical transmission: Direct drive generator

PRODUCTION

Production quantity to date: 5 (June 1980)

MISCELLANEOUS

One year warranty.
Annual maintenance required.

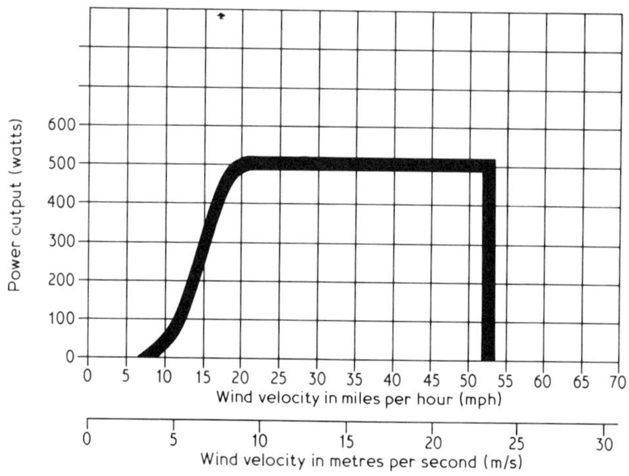

Alternative Energy Ltd,
Woodford, Co. Galway,
Ireland
Telephone: Woodford 52
Telex: 5339

Alternative Energy Ltd 10 kW

PERFORMANCE AND OPERATIONAL DATA

Rated electrical power	10 000 watt
Rated output voltage	380 volt, 3-phase/ 220 volt d.c.
Cut in wind speed	3.6 m/s
Rated wind speed	10.7 m/s
Shut-down wind speed	24.5–26.8 m/s
Survival wind speed (tested to 29.4 m/s)	53.5 m/s
Shaft speed at rated output	120 rev/min
Maximum shaft speed	180 rev/min

PHYSICAL DATA

Turbine
Type: Horizontal axis, upwind with tail vane
Dimensions
diameter 8.23 m

Blades
Construction: Fibreglass moulding on mild-steel shaft
Number: 2

Control: Variable pitch turbine

Generator: d.c. generator self-excited and cumulative compounded

Mechanical transmission: Triplex chain drive in oil bath

System weight
615 kg (head unit)
820 kg (tower)

PRODUCTION

First production: 1980
Production quantity to date: 7 (May 1981)

MISCELLANEOUS

One year warranty.
Semi-annual maintenance required.
11.3 m tower.

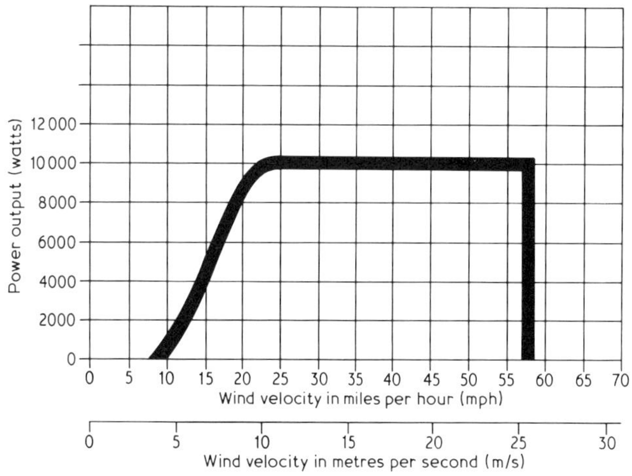

Alternative Energy Ltd,
Woodford, Co. Galway,
Ireland
Telephone: Woodford 52
Telex: 5339

Altos BWP–8B

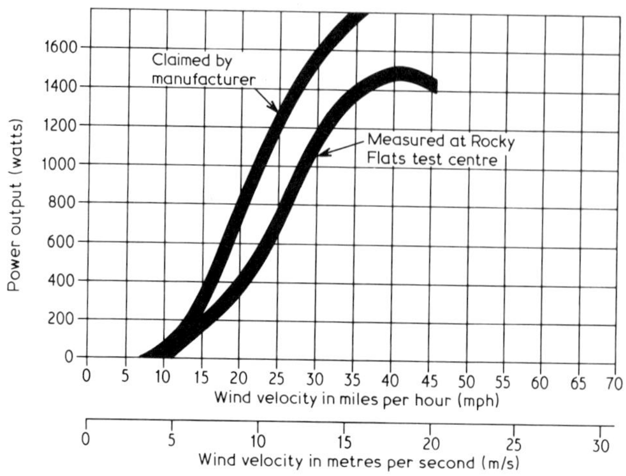

PERFORMANCE AND OPERATIONAL DATA

Rated electrical power	1500 watt
Rated output voltage	24 volt d.c.
Cut in wind speed	4.5 m/s
Rated wind speed	12.5 m/s
Shut-down wind speed	33.5 m/s
Survival wind speed	44.6 m/s
Shaft speed at rated output	165 rev/min
Maximum shaft speed	300 rev/min

PHYSICAL DATA

Turbine

Type: Horizontal axis, upwind with tail vane

Dimensions
diameter 2.32 m

Blades

Construction: Hollow aluminium 5052 blades with lightweight outer rim

Number: 24

Control: Fixed pitch; high thrust turns turbine sideways to the wind

Generator: a.c. generator

Mechanical transmission: 11:1 ratio speed increasing gearbox

System weight
net 114 kg

PRODUCTION

First production: 1975

MISCELLANEOUS

Warranty on turbine, 90 days, covering parts and workmanship. Remainder of major components, one year, parts and workmanship.
Semi-annual fluid and system check; annual lubrication.
Handbook available.

REPORTS

Rocky Flats test centre reports:

Higashi K.K. *Altos–Model 8B Wind Turbine Generator Report* RFP 3033/3533/79–4, July 1979.

Higashi, K.K. and Carr, M.J. *Altos – Model 8B Wind Turbine Generator Failure Analysis* RFP 3035/3533/79–10, 18 June 1979.

Rocky Flats Performance summary sheet, Altos Model 8B, August 1980.

Altos,
PO Box 905, Boulder,
Colorado 80302
USA
Telephone: (303) 443–5422

Altos
BWP–12B

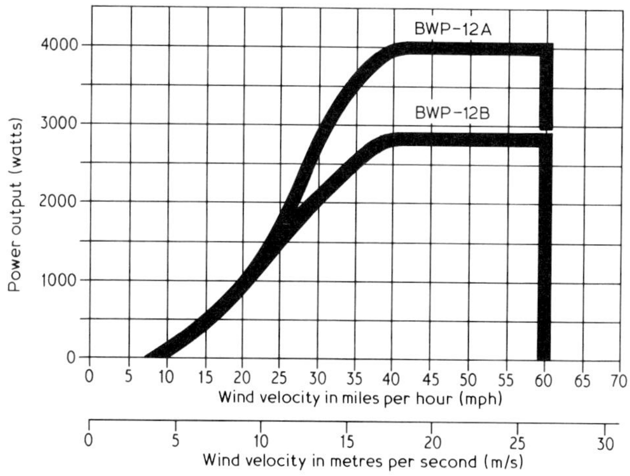

PERFORMANCE AND OPERATIONAL DATA

Rated electrical power	2000 watt
Rated output voltage	24 volt d.c.
Cut-in wind speed	3.6 m/s
Rated wind speed	12.5 m/s
Shut-down wind speed	26.8 m/s
Survival wind speed	40.1 m/s
Shaft speed at rated output	116 rev/min
Maximum shaft speed	200 rev/min

PHYSICAL DATA

Turbine
Type: Horizontal axis, upwind with tail vane
Dimensions
diameter 3.51 m
Blades
Construction: Hollow aluminium 6061 blades with lightweight outer rim
Number: 24
Control: Fixed pitch; high thrust turns turbine sideways to the wind
Generator: a.c. generator
Mechanical transmission: 11:1 ratio speed increasing gearbox
System weight
net 136 kg

PRODUCTION

First production: 1977/8

MISCELLANEOUS

Warranty on turbine, 90 days, covering parts and workmanship. Remainder of major components, one year, parts and workmanship.
Semi-annual fluid and system check; annual lubrication.
Handbook available.

VARIANTS

Model BWP–12A, which has unrectified generator output, rated at 2200 watt, 115/200 volt, 3 phase. Otherwise the same except a 17:1 ratio gearbox.

Altos,
PO Box 905, Boulder,
Colorado 80302
USA
Telephone: (303) 443–5422

Reinke Wind Turbine

PERFORMANCE AND OPERATIONAL DATA

Rated electrical power	7500 watt
Rated output voltage	208 volt
Cut-in wind speed	~2.5 m/s
Rated wind speed	~15.5 m/s
Shut-down wind speed	17.8 m/s
Survival wind speed	62.4 m/s
Shaft speed at rated output	220 rev/min
Maximum shaft speed	240–250 rev/min

PHYSICAL DATA

Turbine
Type: Horizontal axis, upwind with tail vane
Dimensions
diameter 5.80 m
Blades
Construction: Aluminium
Number: 3
Control: Variable pitch turbine
Generator: 26 pole, 3-phase permanent magnet, a.c. generator, rated 10 kVA, 400 Hz
Mechanical transmission: Speed increasing belt drive
System weight: 1100 kg (net)

PRODUCTION

First production: 1978
Production quantity to date: 25 (May 1981)

MISCELLANEOUS

16.5 m lattice tower supplied.
Annual maintenance requirement.
One year warranty.

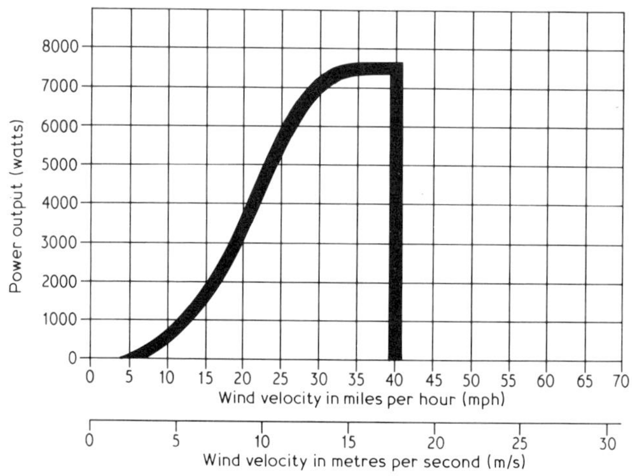

American Energy Savers,
912 St. Paul Road,
PO Box 1421,
Grand Island,
Nebraska 68801,
USA

American Wind Turbine AWP–8

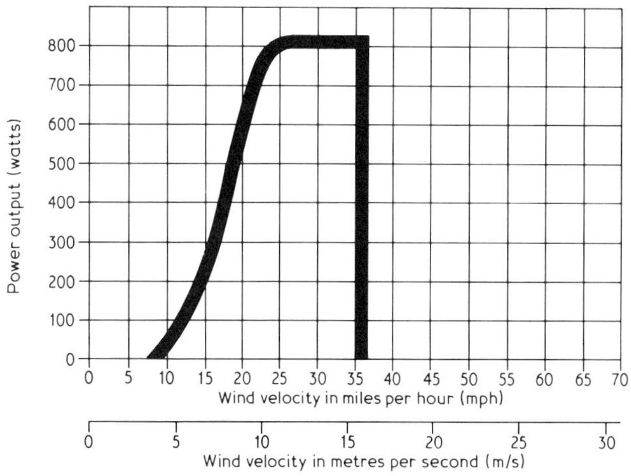

PERFORMANCE AND OPERATIONAL DATA

Rated electrical power	500 watt
Rated output voltage	12/24/28/32 volt d.c.
Cut-in wind speed	4.5 m/s
Rated wind speed	8.9 m/s
Shut-down wind speed	15.6 m/s
Survival wind speed	53.5 m/s
Shaft speed at rated output	150 rev/min

PHYSICAL DATA

Turbine
Type: Horizontal axis, upwind with tail vane
Dimensions
diameter 2.33 m
Blades
Construction: Aluminium with lightweight outer rim
Number: 24
Control: Fixed pitch; high thrust turns rotor sideways to the wind
Generator: Permanent magnet a.c. generator (rectified)
Mechanical transmission: Belt drive from rim, 30:1 effective ratio

PRODUCTION

First production: Late 1977

MISCELLANEOUS

90 day warranty, covering parts and workmanship.
Semi-annual maintenance, check bolts; annual maintenance, grease and inspect system.
Handbook supplied with equipment.

American Wind Turbine Co. Inc,
1016 E. Airport Road,
Stillwater, Oklahoma 74074,
USA
Telephone: (405) 377–5333

American Wind Turbine AWP–12

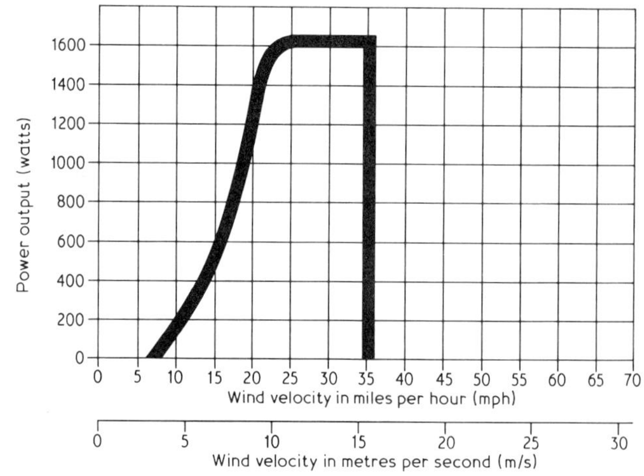

PERFORMANCE AND OPERATIONAL DATA

Rated electrical power	1000 watt
Rated output voltage	12/24/28/32 volt d.c.
Cut-in wind speed	4.5 m/s
Rated wind speed	8.9 m/s
Shut-down wind speed	15.6 m/s
Survival wind speed	44.6 m/s
Shaft speed at rated output	120 rev/min

PHYSICAL DATA

Turbine
Type: Horizontal axis, upwind with tail vane
Dimensions
diameter 3.51 m
Blades
Construction: Aluminium with lightweight outer rim
Number: 36
Control: Fixed pitch; high thrust turns turbine sideways to the wind
Generator: Permanent magnet a.c. generator (rectified)
Mechanical transmission: Belt drive from rim, 30:1 effective ratio
System weight: 145 kg

PRODUCTION

First production: Late 1977

MISCELLANEOUS

90 day warranty, covering parts and workmanship.
Semi-annual maintenance, check bolts; annual maintenance, grease and inspect system.
Handbook supplied with equipment.

American Wind Turbine Co. Inc,
1016 E. Airport Road,
Stillwater, Oklahoma 74074,
USA
Telephone: (405) 377–5333

American Wind Turbine AWP–16

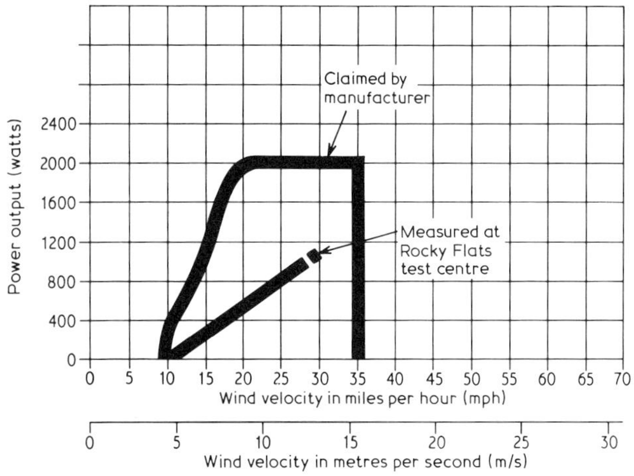

PERFORMANCE AND OPERATIONAL DATA

Rated electrical power	2000 watt
Rated output voltage	12/24/28/32 volt d.c.
Cut-in wind speed	4.5 m/s
Rated wind speed	8.9 m/s
Shut-down wind speed	15.6 m/s
Survival wind speed	44.6 m/s
Maximum shaft speed	90 rev/min

PHYSICAL DATA

Turbine
Type: Horizontal axis, upwind with tail vane
Dimensions
diameter 4.67 m
Blades
Construction: Aluminium with lightweight outer rim
Number: 48
Control: Fixed pitch; high thrust turns turbine sideways to the wind
Generator: 3-phase permanent magnet a.c. generator (rectified)
Mechanical transmission: Driven by belt drive from rim, 30:1 effective ratio
System weight: 190 kg

PRODUCTION

First production: Late 1977

MISCELLANEOUS

90 day warranty, covering parts and workmanship.
Semi-annual maintenance, check bolts; annual maintenance, grease and inspect system.
Handbook supplied with equipment.

REPORTS

Rocky Flats Test Centre Reports:
Higashi, K.K. *American Wind Turbine Model AWP–16, Final Test Report* RFP 3263, March 1980.
Rocky Flats Performance Summary Sheet, American Wind Turbine AWP–16, October 1980.

American Wind Turbine Co. Inc,
1016 E. Airport Road,
Stillwater, Oklahoma 74074,
USA
Telephone: (405) 377–5333

Astral Wilcon AW 10–C

PERFORMANCE AND OPERATIONAL DATA

Rated electrical power	10 000 watt
Rated output voltage	240/480 volt a.c.
(operated in parallel with 3-phase 60 Hz network)	
Cut-in wind speed	4.9 m/s
Rated wind speed	11.16 m/s
Shut-down wind speed	22.3 m/s
Survival wind speed	49.1 m/s
Shaft speed at rated output	131 rev/min
Maximum shaft speed	151 rev/min

PHYSICAL DATA

Turbine
Type: Horizontal axis, upwind with tail vane
Dimensions
diameter 7.94 m
Blades
Construction: Glass reinforced composite
Number: 3
Control: Variable pitch turbine
Generator: Rectified 3 phase a.c. brushless alternator
Mechanical transmission: Two-stage helical gearbox
System weight: 530 kg

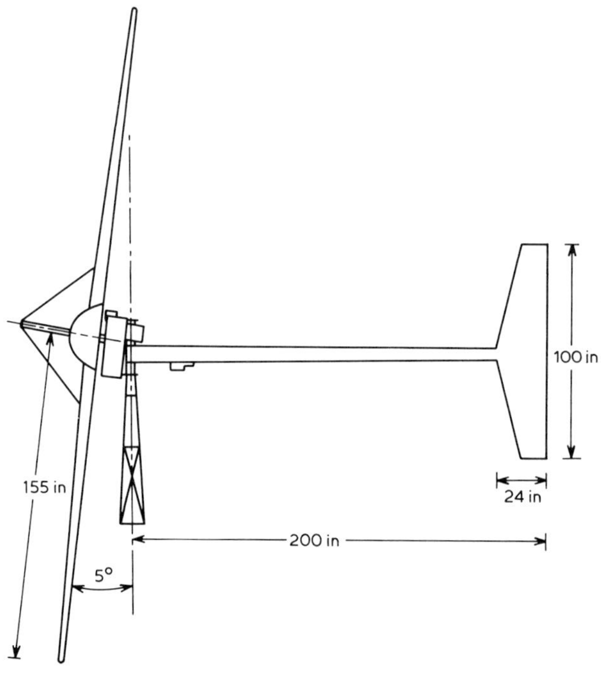

PRODUCTION

First production: October 1979
Production quantity to date: 7 (October 1982)

MISCELLANEOUS

Semi-annual maintenance requirement.
One year warranty.
Blades of various types supplied separately.
Power–wind speed curve for AW10-C not yet available (curve below shows previous model AW 10-B).

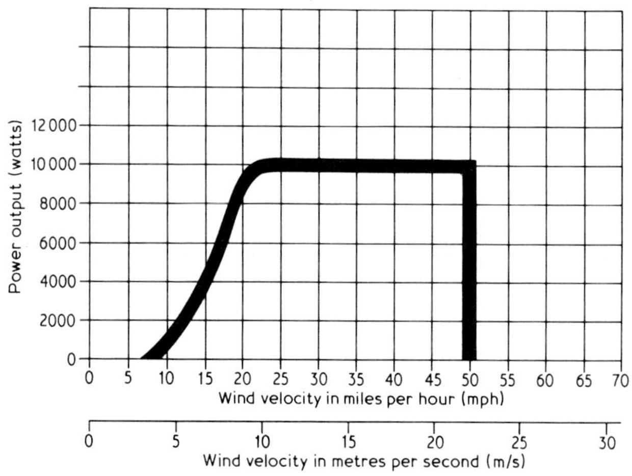

Astral Wilcon,
PO Box 291, Millbury,
Massachusetts 01527,
USA

Bergey Windpower BWC 1000

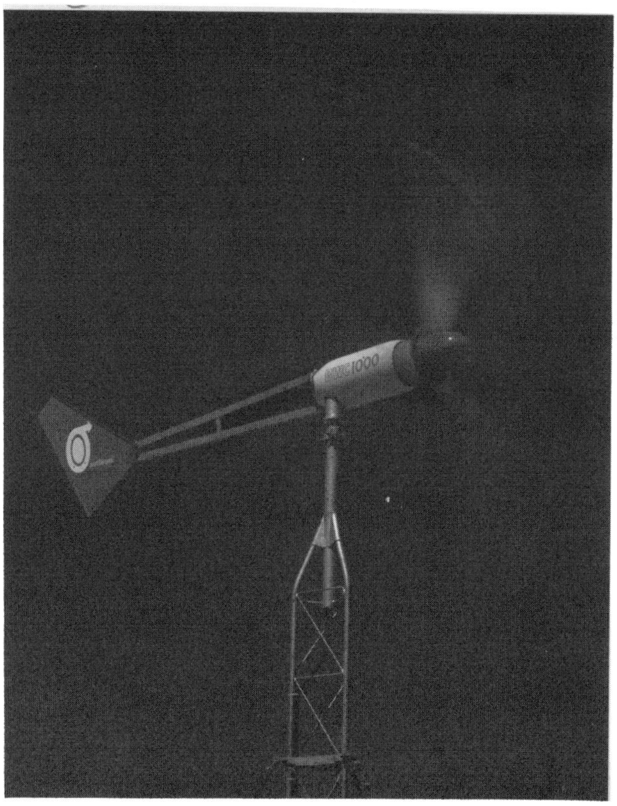

PERFORMANCE AND OPERATIONAL DATA

Rated electrical power	1000 watt
Rated output voltage	12/24/36/48/120 volt d.c.
Cut-in wind speed	4.0 m/s
Rated wind speed	10.7 m/s
Shut-down wind speed	14.3 m/s
Survival wind speed	53.5 m/s
Shaft speed at rated output	620 rev/min

PHYSICAL DATA

Turbine

Type: Horizontal axis, upwind type with tail vane

Dimensions

diameter 2.80 m

Blades

Number: 3

Control: Variable pitch turbine; high thrust turns turbine sideways to wind

Generator: Permanent magnet a.c. generator (rectified)

Mechanical transmission: Direct coupling to turbine

System weight: 55 kg (excluding control system)

PRODUCTION

First production: July 1980

Production quantity to date: 280 (September 1982)

MISCELLANEOUS

No scheduled maintenance or lubrication required.

VARIANTS

Can be supplied with 'Powersync' inverter for operation at 115 volt a.c. in parallel with 60 Hz network (BWC 1000–S model).

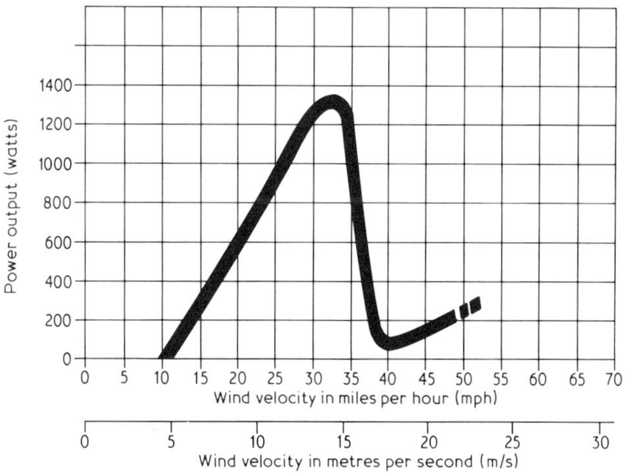

Bergey Windpower Co. Inc,
2001 Priestley Ave,
Norman, Oklahoma 73069,
USA
Telephone: (405) 364–4212

Bertoia AES, Slow-Wind

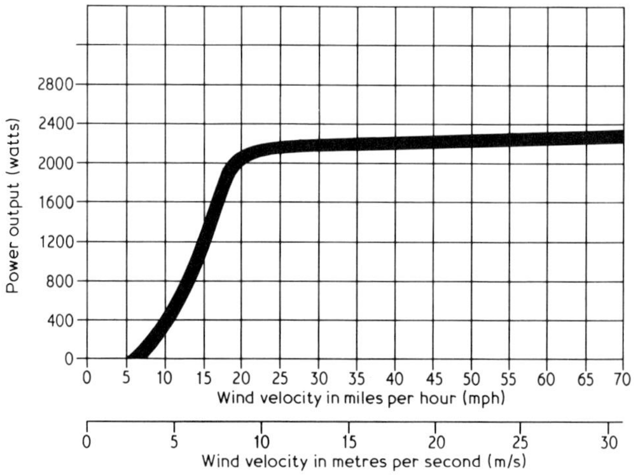

PERFORMANCE AND OPERATIONAL DATA

Rated electrical power	2000 watt
Rated output voltage	120 volt a.c.
Cut-in wind speed	3.3 m/s
Rated wind speed	9.0 m/s
Shut-down wind speed	31.2 m/s
Survival wind speed	40 m/s
Shaft speed at rated output	130 rev/min
Maximum shaft speed	150 rev/min

PHYSICAL DATA

Turbine

Type: Horizontal axis, downwind, operating freely in yaw using turbine coning

Dimensions

diameter 6.0 m

Blades

Construction: 6061–T6 aluminium, creased skin over tube

Number: 3

Control: Variable pitch blades

Generator: Permanent magnet a.c. generator

Mechanical transmission: Direct coupling to turbine

System weight

net 181 kg

PRODUCTION

First production: 1976

Production quantity to date: 7 (September 1982)

MISCELLANEOUS

Maintenance schedule, three monthly inspection of bolts, brushes on sliprings and greasing of bearings.

Three year warranty on complete unit and tower (excluding electrical work).

Handbook available.

Bertoia Studio Ltd,
644 Main Street, Bally,
Pennsylvania 19503,
USA
Telephone: (215) 845 7096

Bristol Aerospace Licus® A

PERFORMANCE AND OPERATIONAL DATA

Rated electrical power	160 watt
Rated output voltage	12/24/48 volt d.c.
Cut-in wind speed	3.1 m/s
Rated wind speed	7.8 m/s
Shut-down wind speed	15.6 m/s
Survival wind speed	44.6 m/s
Shaft speed at rated output	174 rev/min
Maximum shaft speed (nominal)	300 rev/min

PHYSICAL DATA

Turbine
Type: Vertical axis, 'troposkien' type, with 'Savonius' type starting devices
Dimensions
height 3.14 m
diameter 3.14 m
Blades
Construction: Stainless steel
Number: 3
Control: Fixed pitch blades with aerodynamic flaps for overspeed prevention
Generator: Permanent magnet a.c. generator (rectified)
Mechanical transmission: Direct coupling to turbine
System weight
net 195 kg

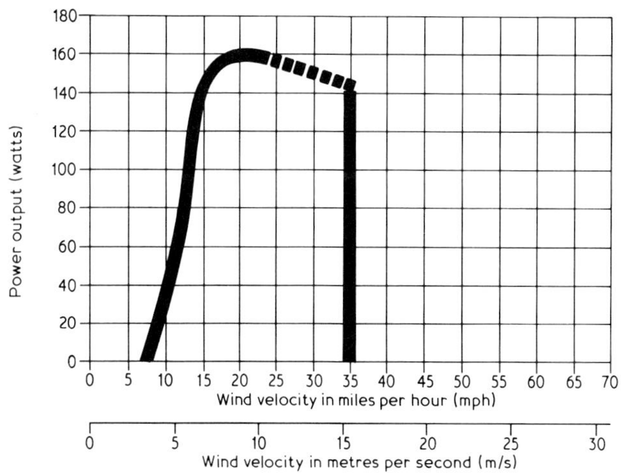

PRODUCTION

First production: 1977
Production quantity to date: 2 (May 1981)

MISCELLANEOUS

Warranty covers complete system for one year.
Maintenance on blades and a.c. generator is required every five years.
Handbook supplied with equipment.
Licus® registered trade mark of Bristol Aerospace Ltd.

Bristol Aerospace Ltd,
PO Box 874, Winnipeg,
Manitoba R3C 2S4,
Canada
Telephone: (204) 775–8331
Telex: 07–57774 to
07–57804

Bristol Aerospace Licus® B

PERFORMANCE AND OPERATIONAL DATA

Rated electrical power	800 watt
Rated output voltage	12/24/48 volt d.c.
Cut-in wind speed	3.1 m/s
Rated wind speed	7.8 m/s
Shut-down wind speed	11.1 m/s
Survival wind speed	49.0 m/s
Shaft speed at rated output	126 rev/min
Maximum shaft speed (nominal)	186 rev/min

PHYSICAL DATA

Turbine

Type: Vertical axis, 'troposkien' type, with 'Savonius' type starting devices

Dimensions

height 6.72 m

diameter 5.60 m

Blades

Construction: Stainless steel made of two piece 'wrap and weld' sheet with integral spar; aerofoil is stretched/bent to troposkien shape

Number: 2

Control: Fixed pitch blades with aerodynamic flaps for overspeed prevention

Generator: Permanent magnet a.c. generator (rectified)

Mechanical transmission: Direct coupling to turbine

System weight

net 726 kg

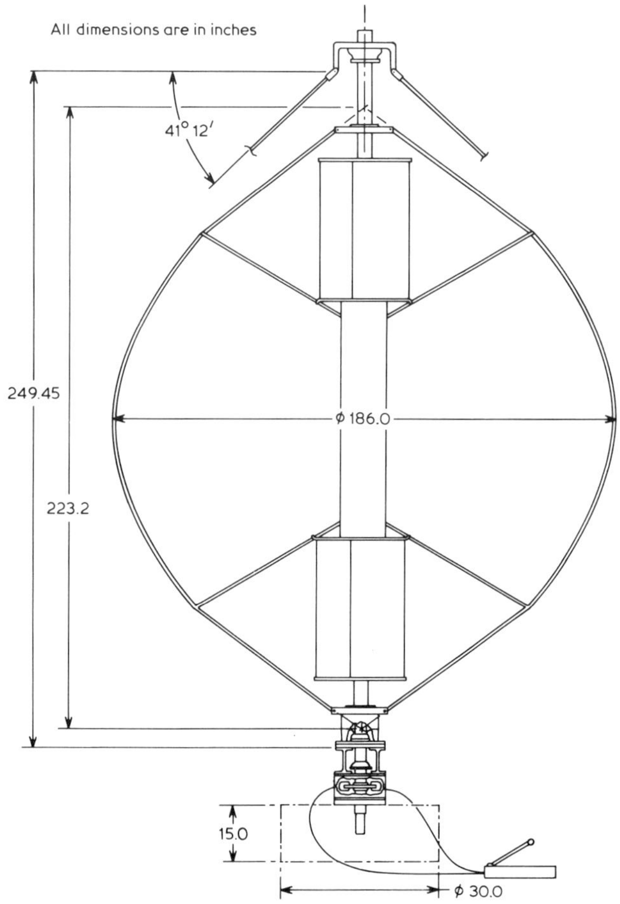

PRODUCTION

First production: 1978

Production quantity to date: 2 (May 1981)

MISCELLANEOUS

Warranty covers complete system for one year.

Maintenance on blades and a.c. generator is required every five years.

Handbook supplied with equipment.

Licus® registered trade mark of Bristol Aerospace Ltd.

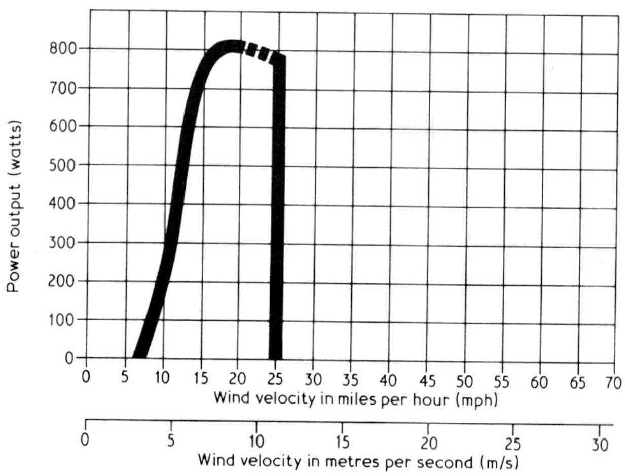

Bristol Aerospace Ltd,
PO Box 874, Winnipeg,
Manitoba R3C 2S4,
Canada
Telephone: (204) 775–8331
Telex: 07–57774 to
07–57804

Bristol Aerospace Licus® C

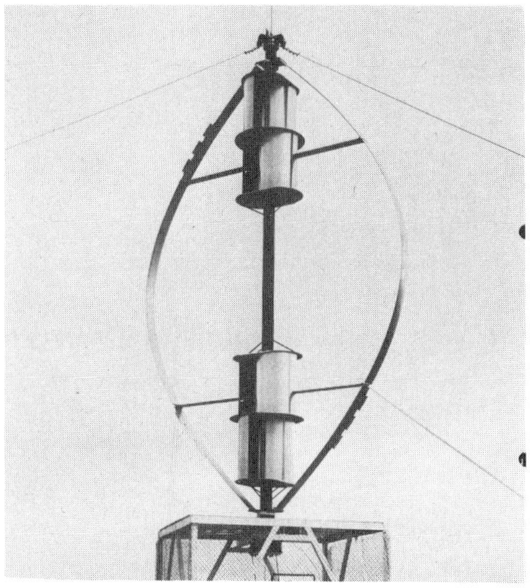

PERFORMANCE AND OPERATIONAL DATA

Rated electrical power	3500 watt
Rated output voltage	24/48/96 volt d.c.
Cut-in wind speed	3.6 m/s
Rated wind speed	7.7 m/s
Shut-down wind speed	11.1 m/s
Survival wind speed	49.0 m/s
Shaft speed at rated output	110 rev/min
Maximum shaft speed (nominal)	175 rev/min

PHYSICAL DATA

Turbine
Type: Vertical axis, 'troposkien' type, with 'Savonius' type starting devices
Dimensions
height 11.52 m
diameter 7.2 m
Blades
Construction: Stainless steel made of two piece 'wrap and weld' sheet with integral spar; aerofoil is stretched/bent to troposkien shape
Number: 2
Control: Fixed pitch blades with aerodynamic flaps for overspeed
Generator: Permanent magnet a.c. generator (rectified)
Mechanical transmission: Direct coupling to turbine
System weight
net 1407 kg

PRODUCTION

First production: 1978
Production quantity to date: 4 (May 1981)

MISCELLANEOUS

Warranty covers complete system for one year.
Maintenance on blades and a.c. generator is required every five years.
Handbook supplied with equipment.
Licus® registered trade mark of Bristol Aerospace Ltd.

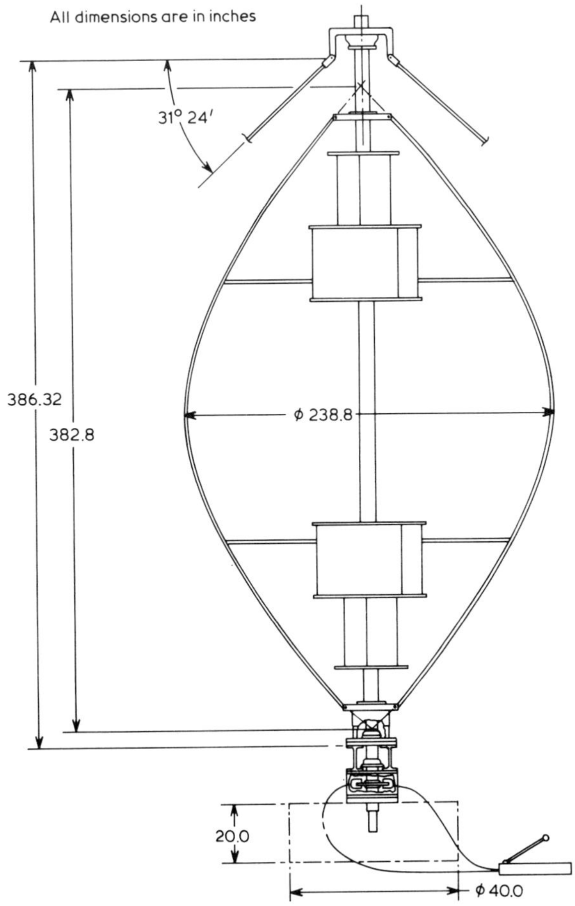

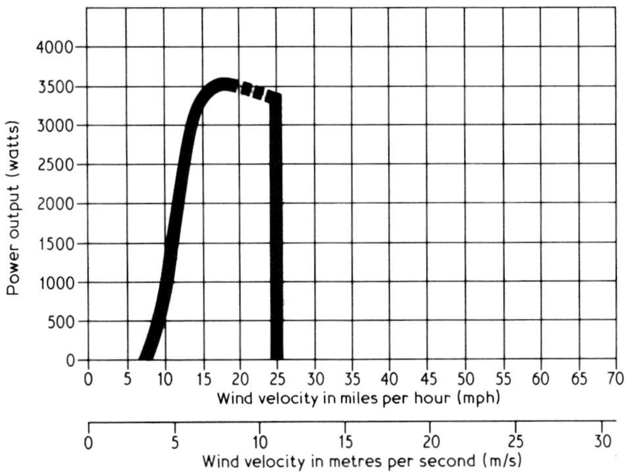

Bristol Aerospace Ltd,
PO Box 874, Winnipeg,
Manitoba R3C 2S4,
Canada
Telephone: (204) 775–8331
Telex: 07–57774 to
07–57804

Hermann Brümmer Windkraftlagen KG BW 51

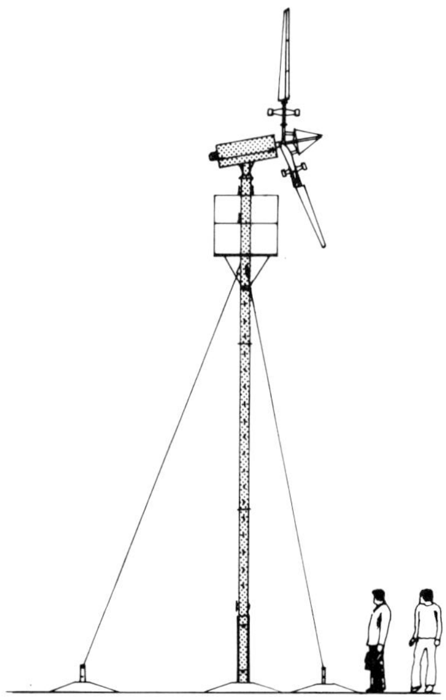

PERFORMANCE AND OPERATIONAL DATA

Rated electrical power	2000 watt
Rated output voltage	380 volt a.c. 3 phase, 50 Hz
Cut-in wind speed	3 m/s
Rated wind speed	8 m/s
Shut-down wind speed	25 m/s
Survival wind speed	40 m/s
Shaft speed at rated output	60 rev/min
Maximum shaft speed	66 rev/min

PHYSICAL DATA

Turbine
Type: Horizontal axis, downwind operating freely in yaw
Dimensions
diameter 5 m

Blades
Construction: Sheet steel with ribs and spars
Number: 3

Control: Centrifugally operated blade pitch control for overspeed protection

Generator: Synchronous or induction generator

Mechanical transmission: 25:1 speed increasing two-stage oil bath gearbox

System weight: 280 kg

PRODUCTION

First production: 1969
Production quantity to date: 9 (January 1980)

MISCELLANEOUS

One year warranty.
Annual maintenance comprising lubrication.
9 m guyed tubular tower with working platform and railings is standard.

VARIANTS

Different generators up to double capacity can be incorporated.
Concrete or lattice towers are available, also different tower heights.

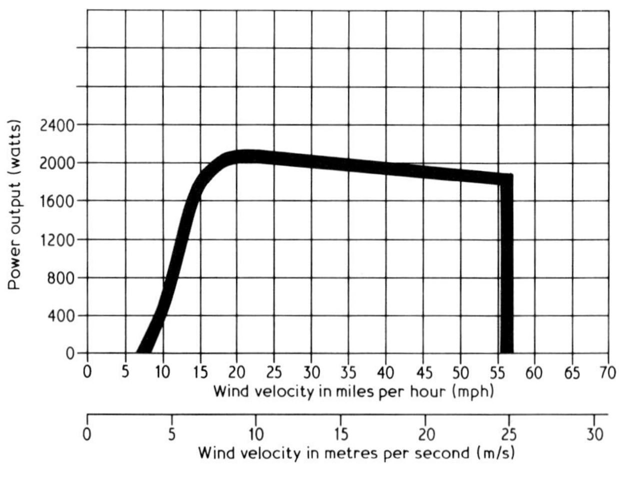

Hermann Brümmer
Windkraftlagen KG,
Mühlenstrasse 1–8,
3522 Bad Karlshafen 2,
West Germany
Telephone: 0 56 72/8 20

Hermann Brümmer Windkraftlagen KG BW 61

PERFORMANCE AND OPERATIONAL DATA

Rated electrical power	3000 watt
Rated output voltage	380 volt a.c. 3 phase, 50 Hz
Cut-in wind speed	3 m/s
Rated wind speed	8 m/s
Shut-down wind speed	25 m/s
Survival wind speed	40 m/s
Shaft speed at rated output	60 rev/min
Maximum shaft speed	66 rev/min

PHYSICAL DATA

Turbine
Type: Horizontal axis, downwind operating freely in yaw
Dimensions
diameter 6 m
Blades
Construction: Sheet steel with ribs and spars
Number: 3
Control: Centrifugally operated blade pitch control for overspeed protection
Generator: Synchronous or induction generator
Mechanical transmission: 25:1 speed increasing two-stage oil bath gearbox
System weight: 400 kg

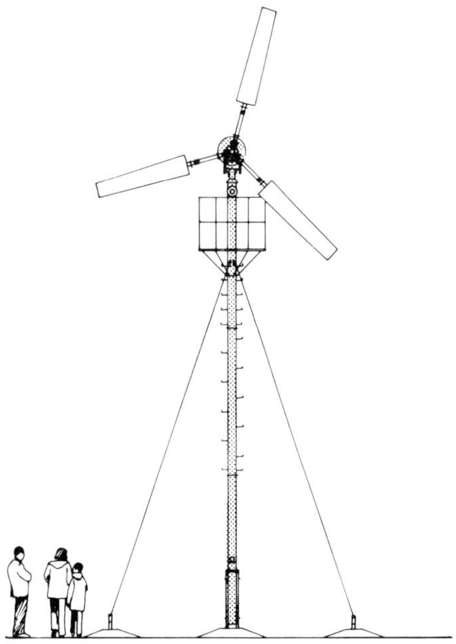

PRODUCTION

First production: 1968
Production quantity to date: 17 (January 1980)

MISCELLANEOUS

One year warranty.
Annual maintenance comprising lubrication.
9 m guyed tubular tower with working platform and railings is standard.

VARIANTS

Different generators up to double capacity can be incorporated.
Concrete or lattice towers are available, also different tower heights.

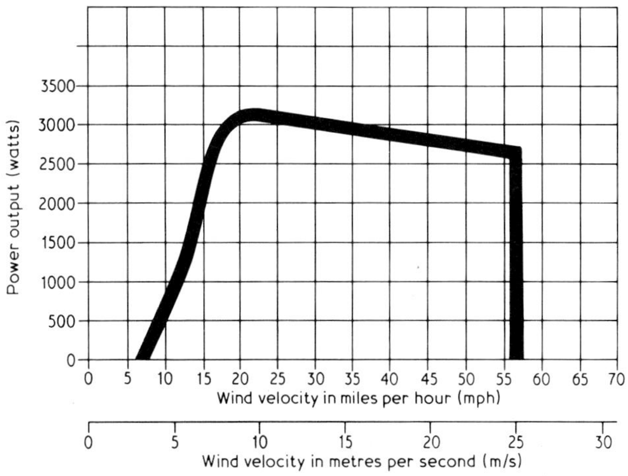

Hermann Brümmer
Windkraftlagen KG,
Mühlenstrasse 1–8,
3522 Bad Karlshafen 2,
West Germany
Telephone: 0 56 72/8 20

Hermann Brümmer Windkraftlagen KG BW 81

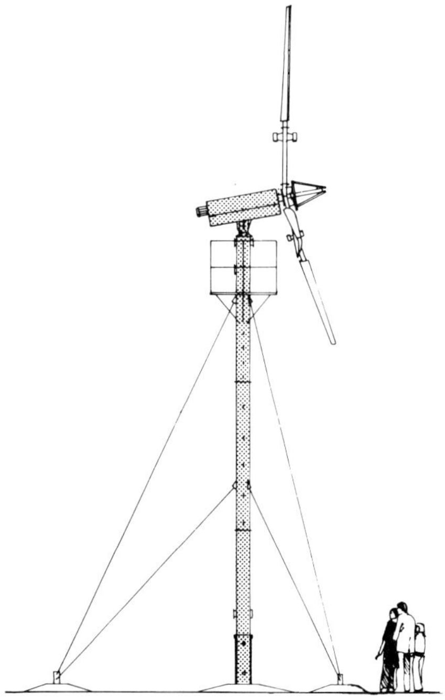

PERFORMANCE AND OPERATIONAL DATA

Rated electrical power	7000 watt
Rated output voltage	380 volt a.c. 3 phase, 50 Hz
Cut-in wind speed	3 m/s
Rated wind speed	8 m/s
Shut-down wind speed	25 m/s
Survival wind speed	40 m/s
Shaft speed at rated output	50 rev/min
Maximum shaft speed	55 rev/min

PHYSICAL DATA

Turbine

Type: Horizontal axis, downwind operating freely in yaw

Dimensions

diameter 8 m

Blades

Construction: Sheet steel with ribs and spars

Number: 3

Control: Centrifugally operated blade pitch control for overspeed protection

Generator: Synchronous or induction generator

Mechanical transmission: 30:1 speed increasing two-stage oil bath gearbox

System weight: 700 kg

PRODUCTION

First production: 1971

Production quantity to date: 23 (January 1980)

MISCELLANEOUS

One year warranty.

Annual maintenance comprising lubrication.

9 m guyed tubular tower with working platform and railings is standard.

VARIANTS

Different generators up to double capacity can be incorporated.

12 m guyed tower or lattice or concrete towers of different heights are available.

Hermann Brümmer
Windkraftlagen KG,
Mühlenstrasse 1–8,
3522 Bad Karlshafen 2,
West Germany
Telephone: 0 56 72/8 20

Hermann Brümmer Windkraftlagen KG BW 120

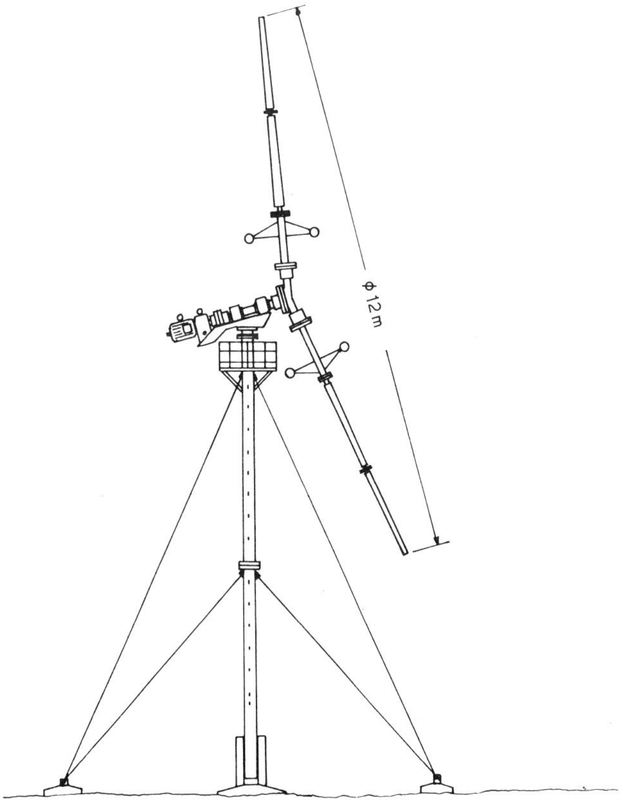

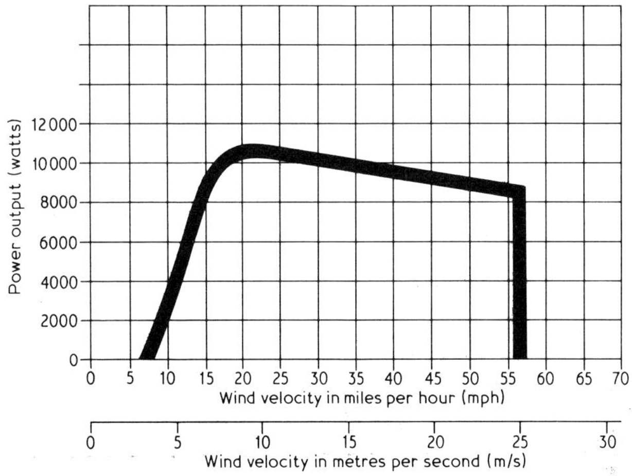

PERFORMANCE AND OPERATIONAL DATA

Rated electrical power	10 000 watt
Rated output voltage	380 volt a.c. 3 phase, 50 Hz
Cut-in wind speed	3 m/s
Rated wind speed	8 m/s
Shut-down wind speed	25 m/s
Survival wind speed	40 m/s
Shaft speed at rated output	40 rev/min
Maximum shaft speed	44 rev/min

PHYSICAL DATA

Turbine
Type: Horizontal axis, downwind operating freely in yaw
Dimensions
diameter 12 m
Blades
Construction: Sheet steel with ribs and spars
Number: 3 (each with 2 segments)
Control: Centrifugally operated blade pitch control for overspeed protection
Generator: Synchronous or induction generator
Mechanical transmission: 37.5:1 speed increasing two-stage oil bath gearbox
System weight: 1200 kg

PRODUCTION

First production: 1974
Production quantity to date: 6 (January 1980)

MISCELLANEOUS

One year warranty.
Annual maintenance comprising lubrication.
9 m guyed tubular tower with working platform and railings is standard.

VARIANTS

Different generators up to double capacity can be incorporated.
12 m guyed tower or lattice or concrete towers of different heights are available.

Hermann Brümmer
Windkraftlagen KG,
Mühlenstrasse 1–8,
3522 Bad Karlshafen 2,
West Germany
Telephone: 0 56 72/8 20

Hermann Brümmer Windkraftlagen KG BW 150

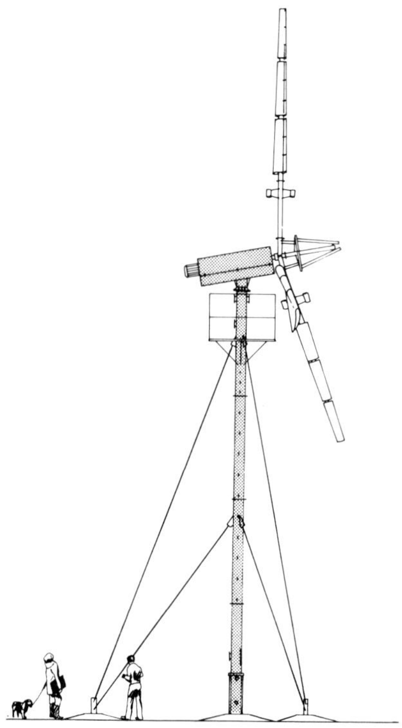

PERFORMANCE AND OPERATIONAL DATA

Rated electrical power	15 000 watt
Rated output voltage	380 volt a.c. 3 phase, 50 Hz
Cut-in wind speed	3 m/s
Rated wind speed	8 m/s
Shut-down wind speed	25 m/s
Survival wind speed	40 m/s
Shaft speed at rated output	35 rev/min
Maximum shaft speed	38 rev/min

PHYSICAL DATA

Turbine

Type: Horizontal axis, downwind operating freely in yaw

Dimensions

diameter 15 m

Blades

Construction: Sheet steel with ribs and spars

Number: 3 (each with 3 segments)

Control: Centrifugally operated blade pitch control for overspeed protection

Generator: Synchronous or induction generator

Mechanical transmission: 42:1 speed increasing two-stage oil bath gearbox

System weight: 2000 kg

PRODUCTION

First production: 1977

Production quantity to date: 3 (January 1980)

MISCELLANEOUS

One year warranty.

Annual maintenance comprising lubrication.

12 m guyed tubular tower with working platform and railings is standard.

VARIANTS

Different generators up to double capacity can be incorporated.

15 m guyed tower or lattice or concrete towers of different heights are available.

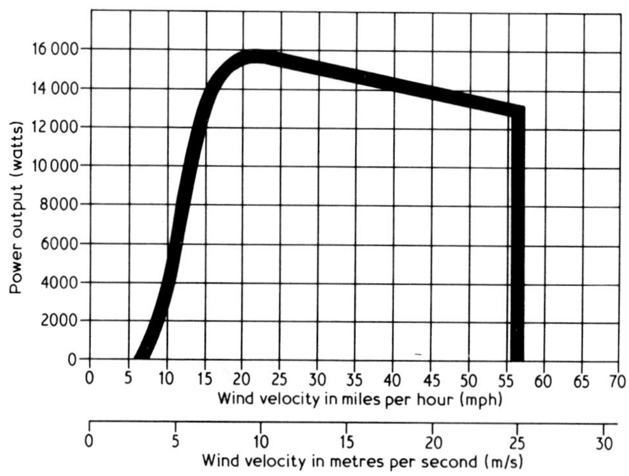

Hermann Brümmer
Windkraftlagen KG,
Mühlenstrasse 1–8,
3522 Bad Karlshafen 2,
West Germany
Telephone: 0 56 72/8 20

DAF INDAL 37×55 VAWT

PERFORMANCE AND OPERATIONAL DATA

Rated electrical power	56 000 watt
Rated output voltage	460/575 volt a.c. 3 phase, 60 Hz
Cut-in wind speed	5.8 m/s
Rated wind speed	17.3 m/s
Shut-down wind speed	20 m/s
Survival wind speed	65 m/s
Shaft speed at rated output	80 rev/min
Maximum shaft speed	110 rev/min

PHYSICAL DATA

Turbine
Type: Vertical axis 'troposkien' type
Dimensions
height 16.79 m
diameter 11.2 m

Blades
Construction: Extruded aluminium
Number: 2

Control: Fixed pitch blades with optional aerodynamic flaps for overspeed prevention

Generator: Induction generator

Mechanical transmission: Speed increasing gearbox

System weight
net 3630 kg

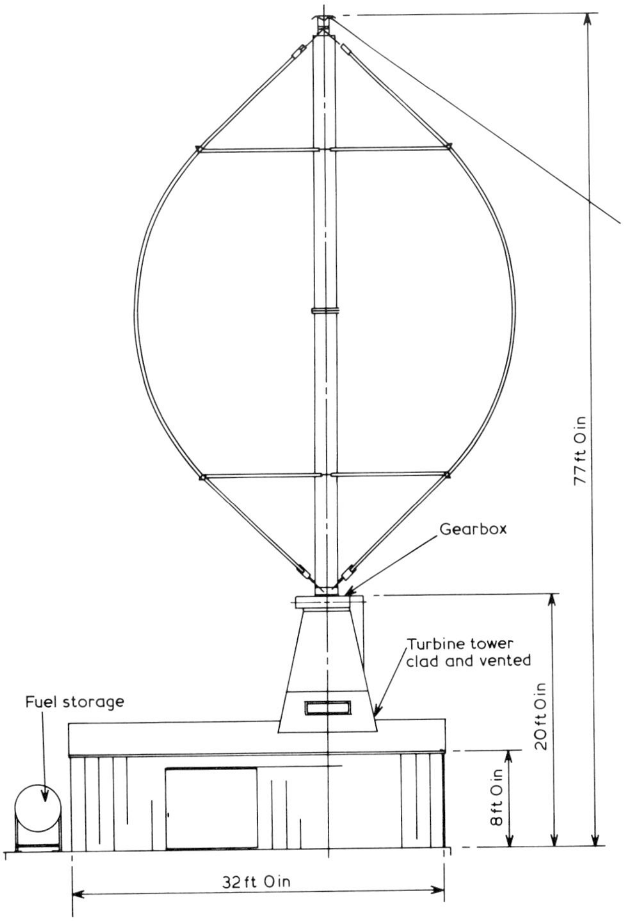

PRODUCTION

First production: 1978
Production quantity to date: 12 (October 1982)

MISCELLANEOUS

Handbook available.
Machine operates in parallel with the network at constant speed set by network frequency.

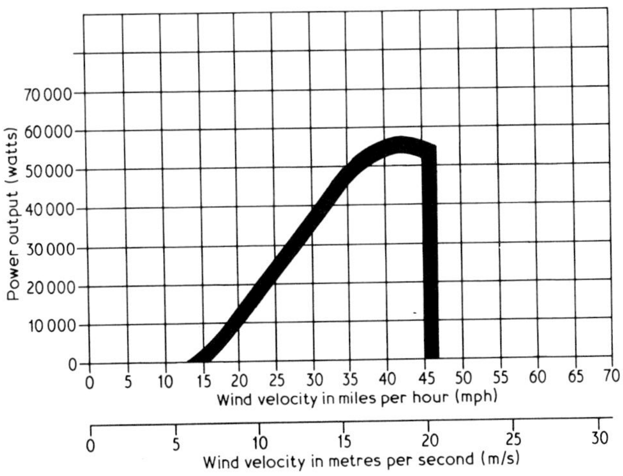

DAF INDAL Ltd,
3570 Hawkestone Road,
Mississauga, Ontario L5C 2V8,
Canada
Telephone: (416) 275–5300
Telex: 06–961482

Dakota Wind & Sun BC4

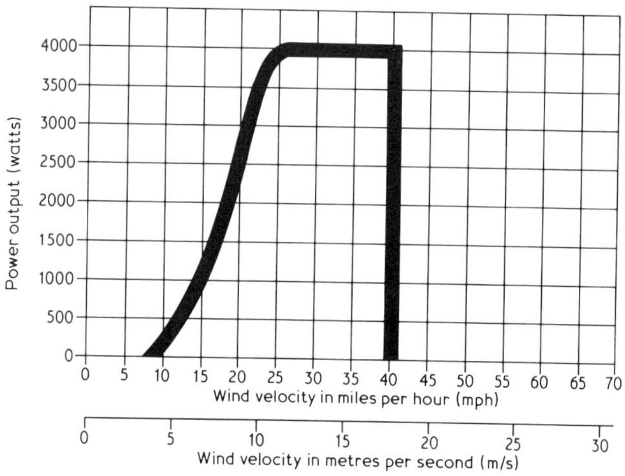

PERFORMANCE AND OPERATIONAL DATA

Rated electrical power	4000 watt
Rated output voltage	140 volt d.c.
Cut-in wind speed	3.6 m/s
Rated wind speed	12.0 m/s
Shut-down wind speed	17.8 m/s
Survival wind speed	40.1 m/s
Shaft speed at rated output	300 rev/min
Maximum shaft speed	325 rev/min

PHYSICAL DATA

Turbine

Type: Horizontal axis, upwind with tail vane
Dimensions
diameter 4.27 m

Blades

Construction: Sitka spruce, epoxy painted with brass leading edge
Number: 3

Control: Centrifugally operated blade pitch control
Generator: d.c. generator
Mechanical transmission: Direct coupling to turbine
System weight
net 275 kg

PRODUCTION

First production: 1977
Production quantity to date: About 110 (September 1979)

MISCELLANEOUS

Limited warranty, covering replacement of defective parts for two years.
Semi-annual maintenance, comprising inspection and lubrication of system.
Guyed (45G–17.1 or 23.2 or 29.3 m) or self-supporting (SSV–18.3 or 24.4 or 30.5 m) Rohn towers, or Valmont steel poles (15.3 or 21.3 or 30.5 m) are offered.

VARIANT

SI 4 model with 200 volt d.c. generator, designed for use with synchronous inverter for network operation (BC 4 model is intended for battery charging).

Dakota Wind & Sun Ltd,
PO Box 1781,
Aberdeen,
South Dakota 57401,
USA
Telephone: (605) 299 0815

Smedermester–Møllen

PERFORMANCE AND OPERATIONAL DATA

Rated electrical power	22 000 watt
Rated output voltage	380 volt a.c.
(operated in parallel with 3-phase 50 Hz network)	
Cut-in wind speed	4.5 m/s
Rated wind speed	12 m/s
Shaft speed at rated output	73 rev/min
Maximum shaft speed	73 rev/min

PHYSICAL DATA

Turbine
Type: Horizontal axis, upwind type with twin fan-tails
Dimensions
diameter 10 m
Blades
Construction: Polyester/fibreglass
Number: 3
Control: Fixed pitch turbine with rotating tip airbrakes
Generator: Induction generator
Mechanical transmission: Two-stage speed increasing helical gearbox
System weight: 2200 kg

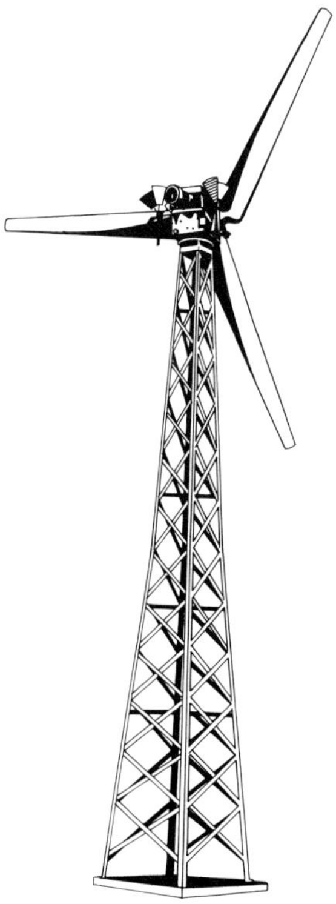

PRODUCTION

First production: 1979
Production quantity to date: 24 (September 1982)

MISCELLANEOUS

One year warranty.
Six monthly maintenance and inspection.

VARIANTS

Different tower heights offered.
A two generator version is available.

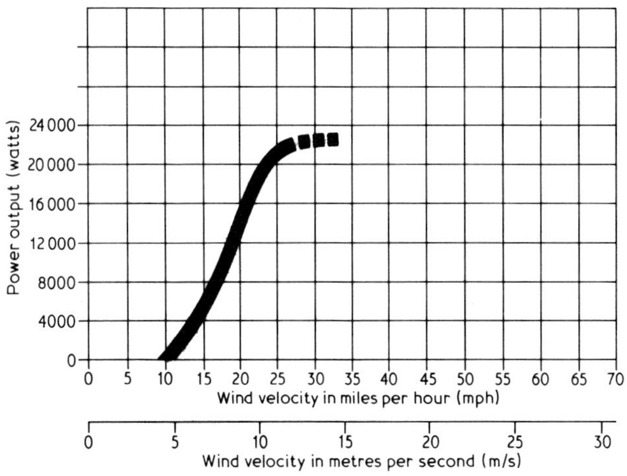

Dansk Smedermester-forening,
Sdr. Boulevard 83,
5000 Odense C,
Denmark
Telephone: (09) 14 1330

Dunlite
1 kW

PERFORMANCE AND OPERATIONAL DATA

Rated electrical power	1000 watt
Rated output voltage	12 volt d.c.
Cut-in wind speed	3.1 m/s
Rated wind speed	11.2 m/s
Shut-down wind speed	none
Survival wind speed	36.1 m/s
Shaft speed at rated output	255 rev/min
Maximum shaft speed	260 rev/min

PHYSICAL DATA

Turbine
Type: Horizontal axis, upwind with tail vane
Dimensions
diameter 3.1 m
Blades
Construction: Folded and welded sheet steel on aluminium cast centre
Number: 3
Control: Variable pitch turbine
Generator: Bushless a.c. generator
Mechanical transmission: 5:1 ratio speed increasing gearbox
System weight: 227 kg

PRODUCTION

First production: 1936
Production quantity to date: 10 000 to 12 000

MISCELLANEOUS

Annual maintenance schedule, including check on bolt tightnesses, oil levels etc.
Handbook supplied with each machine.

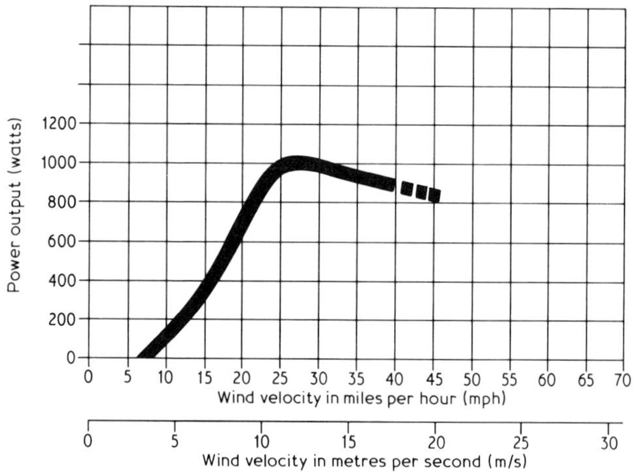

Dunlite,
28 Orsmond Street, Hindmarsh,
South Australia,
Australia 5007
Telephone: (08) 463832
Telex: 82893

Dunlite 2 kW

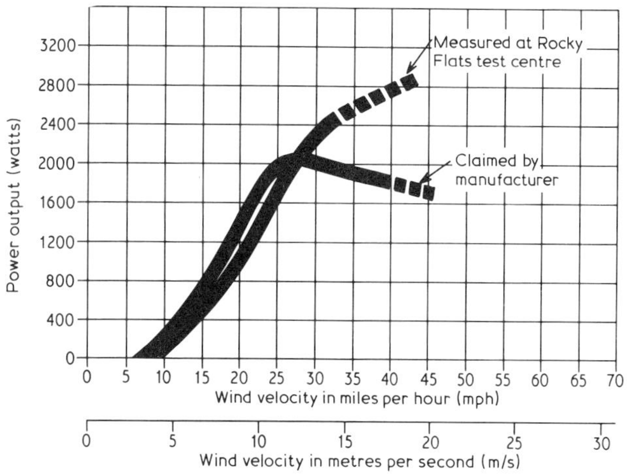

PERFORMANCE AND OPERATIONAL DATA

Rated electrical power	2000 watt
Rated output voltage	24/32/48/110 volt d.c.
Cut-in wind speed	3.1 m/s
Rated wind speed	11.2 m/s
Shut-down wind speed	none
Survival wind speed	36.1 m/s
Shaft speed at rated output	255 rev/min
Maximum shaft speed	260 rev/min

PHYSICAL DATA

Turbine

Type: Horizontal axis, upwind with tail vane

Dimensions

diameter 3.96 m

Blades

Construction: Folded and welded sheet steel on aluminium cast centre

Number: 3

Control: Variable pitch turbine

Generator: Brushless a.c. generator

Mechanical transmission: 5:1 ratio speed increasing gearbox

System weight: 227 kg

PRODUCTION

First production: 1936

Production quantity to date: 10 000 to 12 000

MISCELLANEOUS

Annual maintenance schedule, including check on bolt tightnesses, oil levels etc.

Handbook supplied with each machine.

VARIANTS

'Magnetic latching' system.

REPORTS

Rocky Flats test centre reports:

Higashi, K.K. *Dunlite Model 81/002550, Performance Report* RFP–3026/3533/79–6, July 1979.

Higashi, K.K., Carr, M.J. and Grotsky, V.K. *Dunlite Model 81/002550 Wind Turbine Generator, Failure Analysis* RFP–3028/3533/79–12, 1 June 1979.

Higashi, K.K. *Dunlite Model 81/002550 Wind Turbine Generator, Final Test Report* RFP–3149/3533/80–17, February 1980.

Rocky Flats Performance Summary Sheet, Dunlite Model 81/002550, August 1980.

Dunlite,
28 Orsmond Street, Hindmarsh,
South Australia,
Australia 5007
Telephone: (08) 463832
Telex: 82893

Dunlite
5 kW

PERFORMANCE AND OPERATIONAL DATA

Rated electrical power	5000 watt
Rated output voltage	110 volt d.c.
Cut-in wind speed	3.1 m/s
Rated wind speed	9.5 m/s
Shut-down wind speed	none
Survival wind speed	36.1 m/s
Shaft speed at rated output	220 rev/min
Maximum shaft speed	250 rev/min

PHYSICAL DATA

Turbine
Type: Horizontal axis, upwind with tail vane
Dimensions
diameter 5.6 m
Blades
Construction: Stainless-steel propeller blade arms with cast aluminium hub
Number: 3
Control: Variable pitch turbine
Generator: 8 pole, 3 phase brushless a.c. generator
Mechanical transmission: 5:1 ratio speed increasing gearbox
System weight: 265 kg

MISCELLANEOUS

Annual maintenance schedule, including check on bolt tightnesses, oil levels etc.
Handbook supplied with each machine.

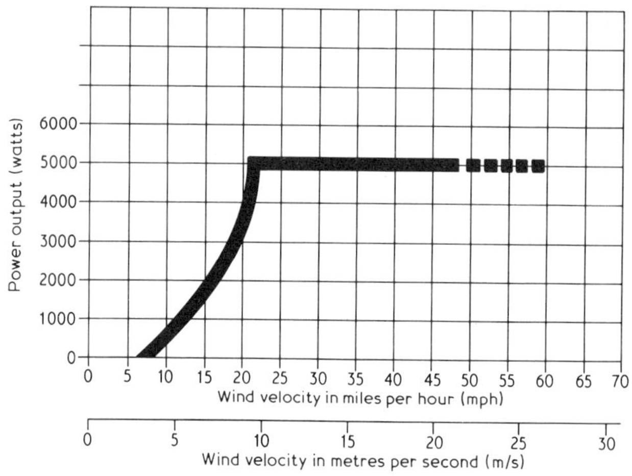

Dunlite,
28 Orsmond Street, Hindmarsh,
South Australia,
Australia 5007
Telephone: (08) 463832
Telex: 82893

Dynergy 5 meter

PERFORMANCE AND OPERATIONAL DATA

Rated electrical power	8000 watt
Rated output voltage	200 volt d.c.
Cut-in wind speed	4.5 m/s
Rated wind speed	10.7 m/s
Shut-down wind speed	20.1 m/s
Survival wind speed	53.5 m/s (parked)
Shaft speed at rated output	240 rev/min
Maximum shaft speed	300 rev/min

PHYSICAL DATA

Turbine
Type: Vertical axis 'troposkien' type
Dimensions
height 4.58 m
diameter 4.58 m
Blades
Construction: Extruded aluminium 6061–T6
Number: 3
Control: Fixed pitch blades
Generator: Permanent magnet a.c. generator
Mechanical transmission: Direct coupling to turbine
System weight:
net 386 kg

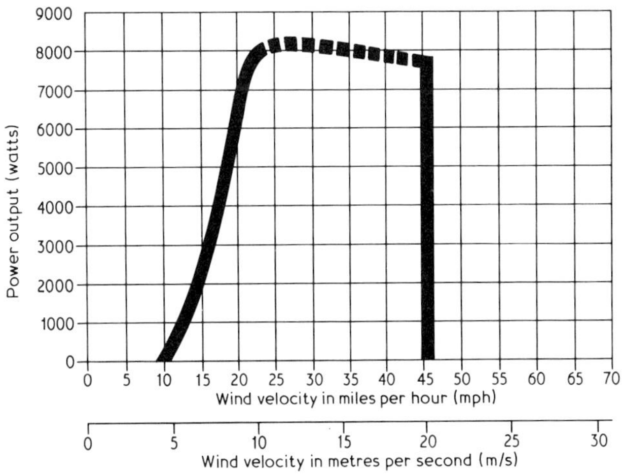

PRODUCTION

First production: 1977
Production quantity to date: About 10 (1979)

MISCELLANEOUS

Three year limited warranty on parts and labour.
Handbook available.
Semi-annual maintenance, comprising system inspection and lubrication.

VARIANTS

A total of 6 variants (systems A to F) are offered. The version detailed (for use with Gemini inverter and network connection) is most common. Also produced is direct a.c. output, variable frequency unit for water or space heating. Other systems offered (but not yet produced) include battery charging, mechanical water pumping (with and without associated batteries) and induction generator for network operation.
Offered without mechanical transmission and generator.

Dynergy Corporation,
PO Box 428,
1269 Union Avenue, Laconia,
New Hampshire 03246
USA
Telephone: (603) 524 8313

Kongstedmøllen

PERFORMANCE AND OPERATIONAL DATA

Rated electrical power	22 000 watt
Rated output voltage	220/380 volt a.c.
(operated in parallel with 3 phase, 50 Hz power network)	
Cut-in wind speed	4 m/s
Survival wind speed	45 m/s
Shaft speed at rated output	60 rev/min
Maximum shaft speed	61 rev/min

PHYSICAL DATA

Turbine

Type: Horizontal axis, upwind with twin fan tails

Dimensions

diameter 10 m

Blades

Construction: Laminated wood (pitch pine) and Bowa veneer

Number: 3

Control: Fixed pitch blades, with rotating blade tips acting as air brakes

Generator: Induction generator

Mechanical transmission: Speed increasing gearbox

System weight

net 800 kg

PRODUCTION

First production: 1979

Production quantity to date: 5 (December 1979)

MISCELLANEOUS

Five year guarantee, including four overhaulings each year.

Egebjerggård
Gammel Gang 2,
Kongsted, 4293 Dianalund
Denmark
Telephone: (03) 56 0213

Elektro W250

PERFORMANCE AND OPERATIONAL DATA

Rated electrical power	150 watt
Rated output voltage	12/24/36 volt d.c.
Cut-in wind speed	*c.* 5 m/s
Rated wind speed	15–20 m/s
Shut-down wind speed	none
Survival wind speed	50–60 m/s
Maximum shaft speed	400 rev/min

PHYSICAL DATA

Turbine
Type: Vertical axis 'Savonius' type
Dimensions
height 1.30 m
diameter 0.66 m
Blades
Construction: Aluminium
Number: 6
Control: Fixed pitch blades
Generator: Permanent magnet a.c. generator (rectified)
Mechanical transmission: Direct coupling to turbine
System weight
net 70 kg
gross 120 kg

PRODUCTION

First production: 1968

MISCELLANEOUS

One-stage automatic load for charging control is available.
101–106 masthead (0.62 m) available for 3 or 4 leg towers, or mast tube only.
Several alternative tower types are available.
Control panel, inverter and batteries also offered.
Annual maintenance comprises lubrication, overall check, paint touch-up (for severe sites and coastal locations twice a year is recommended).
One year warranty according to VSM (Swiss Association of Machinery Manufacturers).

Elektro GmbH,
St Gallerstrasse 27,
Winterthur,
CH–8400, Switzerland
Telephone: (052) 22 34 34
Telex: 76299 Elektro

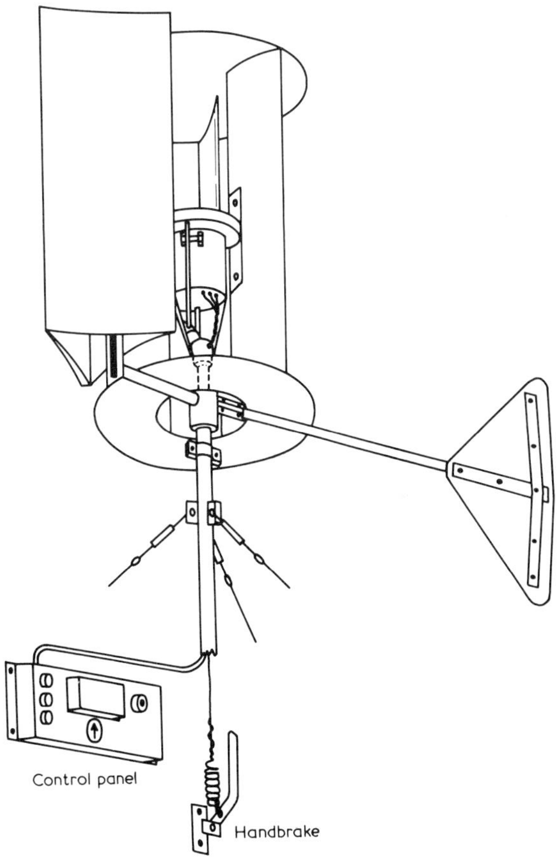

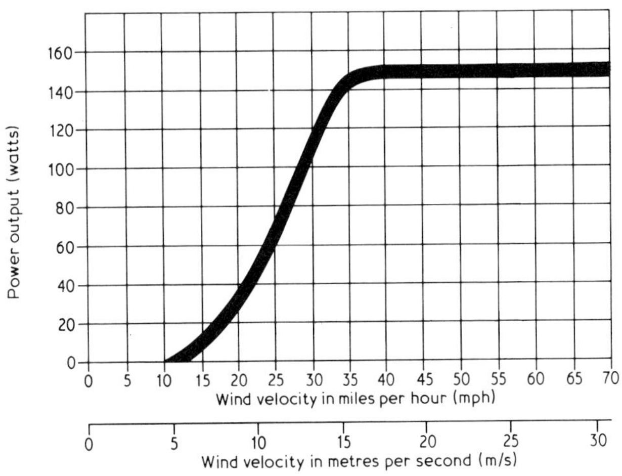

Elektro WV05

PERFORMANCE AND OPERATIONAL DATA

Rated electrical power	500–600 watt
Rated output voltage	12/24/36/48 volt d.c.
Cut-in wind speed	2.5–3 m/s
Rated wind speed	9–10 m/s
Shut-down wind speed	25 m/s
Survival wind speed	60–70 m/s
Maximum shaft speed	700 rev/min

PHYSICAL DATA

Turbine

Type: Horizontal axis, upwind with tail vane
Dimensions
diameter 2.5 m

Blades

Construction: Wood, reinforced with aluminium
Number: 2

Control: Centrifugally operated blade pitch control; signal from auxiliary wind pressure switch used to turn turbine into the plane of the tail vane and the wind

Generator: Permanent magnet a.c. generator (rectified)

Mechanical transmission: Direct coupling to turbine

System weight
net 65 kg
gross 93 kg

PRODUCTION

First production: 1940; earlier types were rated 300/500/800 watt, with a d.c. generator

MISCELLANEOUS

One-stage automatic load for charging control is available.
101–106 masthead (0.62 m) available for 3 or 4 leg towers, or mast tube only.
Several alternative tower types are available.
Control panel, inverter and batteries also offered.
Annual maintenance comprises lubrication, overall check, paint touch-up (for severe sites and coastal locations twice a year is recommended).
One year warranty according to VSM (Swiss Association of Machinery Manufacturers).

Elektro GmbH,
St Gallerstrasse 27,
Winterthur,
CH–8400, Switzerland
Telephone: (052) 22 34 34
Telex: 76299 Elektro

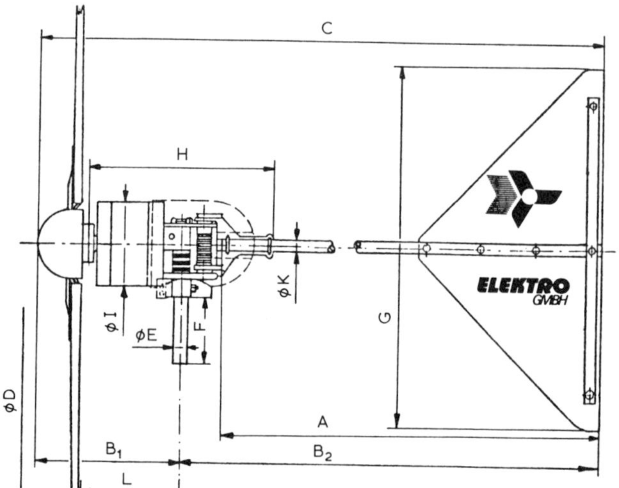

A	1.7	m	F	175	mm
B_1	410	mm	G	950–1000	mm
B_2	1.8	m	H	520	mm
C	2.2	m	ØI	220	mm
ØD	2.5	m	ØK	33.5	mm
D_A	4.9	m²	L	270	mm
ØE	42	mm			

Weight	
Alternator	39 kg
Hub+Propeller	9 kg
Tail	5 kg
Total	53 kg

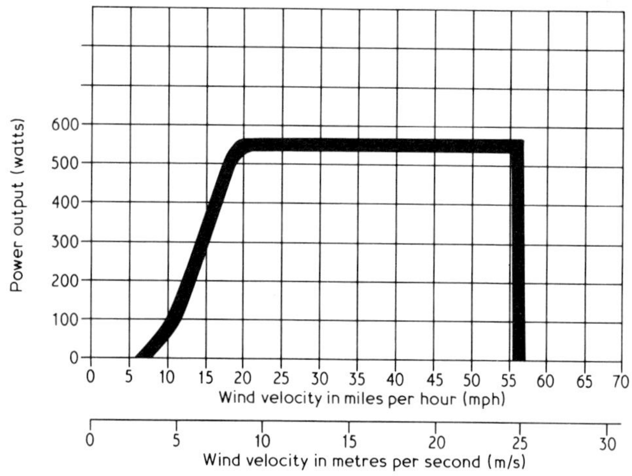

Elektro WV15G

PERFORMANCE AND OPERATIONAL DATA

Rated electrical power	1200 watt
Rated output voltage	12/24/36/48 volt d.c.
Cut-in wind speed	2.5–3 m/s
Rated wind speed	10–11 m/s
Shut-down wind speed	25 m/s
Survival wind speed	70 m/s
Maximum shaft speed	700 rev/min

PHYSICAL DATA

Turbine
Type: Horizontal axis, upwind with tail vane
Dimensions
diameter 3.0 m
Blades
Construction: Wood, reinforced with aluminium
Number: 2
Control: Centrifugally operated blade pitch control; signal from auxiliary wind pressure switch used to turn turbine into the plane of the tail vane and wind
Generator: Permanent magnet a.c. generator (rectified)
Mechanical transmission: Direct coupling to turbine
System weight
net 135 kg
gross 200 kg

PRODUCTION

First production: 1946; earlier types incorporated a d.c. generator

MISCELLANEOUS

Automatic shut-down in storm or with fully charged battery; type 274 unit needs manual restart, type 266 gives automatic restart with 6/12 hour time switch.
111–39a masthead (0.82 m) available for 3 or 4 leg towers, or mast tube only.
Several alternative tower types are available.
Control panel, inverter and batteries also offered.
Annual maintenance comprises lubrication, overall check, paint touch-up (for severe sites and coastal locations twice a year is recommended).
One year warranty according to VSM (Swiss Association of Machinery Manufacturers).

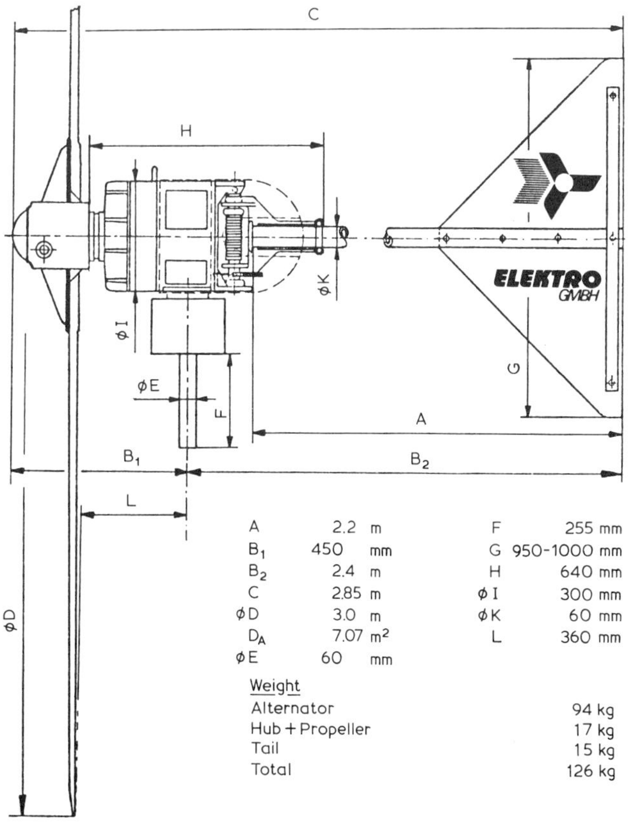

A	2.2 m	F	255 mm
B_1	450 mm	G	950-1000 mm
B_2	2.4 m	H	640 mm
C	2.85 m	ØI	300 mm
ØD	3.0 m	ØK	60 mm
D_A	7.07 m^2	L	360 mm
ØE	60 mm		

Weight	
Alternator	94 kg
Hub + Propeller	17 kg
Tail	15 kg
Total	126 kg

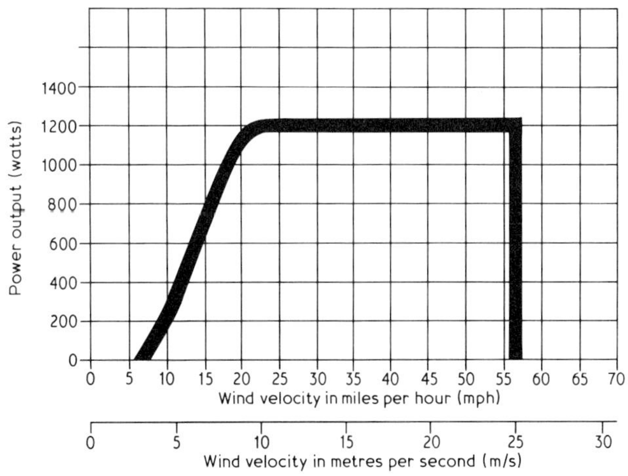

Elektro GmbH,
St Gallerstrasse 27,
Winterthur,
CH–8400, Switzerland
Telephone: (052) 22 34 34
Telex: 76299 Elektro

Elektro WV25G

PERFORMANCE AND OPERATIONAL DATA

Rated electrical power	2200 watt
Rated output voltage	24/36/48/110 volt d.c.
Cut-in wind speed	2.5–3 m/s
Rated wind speed	10–11 m/s
Shut-down wind speed	25 m/s
Survival wind speed	70 m/s
Maximum shaft speed	600 rev/min

PHYSICAL DATA

Turbine
Type: Horizontal axis, upwind with tail vane
Dimensions
diameter 3.6 m
Blades
Construction: Wood, reinforced with aluminium
Number: 2
Control: Centrifugally operated blade pitch control; signal from auxiliary wind pressure switch used to turn turbine into the plane of the tail vane and wind
Generator: Permanent magnet a.c. generator (rectified)
Mechanical transmission: Direct coupling to turbine
System weight
net 180 kg
gross 260 kg

PRODUCTION

First production: 1946; earlier types incorporated a d.c. generator

MISCELLANEOUS

Automatic shut-down in storm or with fully charged battery; type 274 unit needs manual restart, type 266 gives automatic restart with 6/12 hour time switch.
111–39a masthead (0.82 m) available for 3 or 4 leg towers, or mast tube only.
Several alternative tower types available, with mod. 137 assembly crane to suit.
Control panel, inverter and batteries also offered.
Annual maintenance comprises lubrication, overall check, paint touch-up (for severe sites and coastal locations twice a year is recommended).
One year warranty according to VSM (Swiss Association of Machinery Manufacturers).

VARIANTS

WV25D available, with rated output 2000 watt, 110–190 volt a.c. (40–90 Hz) using field rotor a.c. generator for electric heating.

Elektro GmbH,
St Gallerstrasse 27,
Winterthur,
CH–8400, Switzerland
Telephone: (052) 22 34 34
Telex: 76299 Elektro

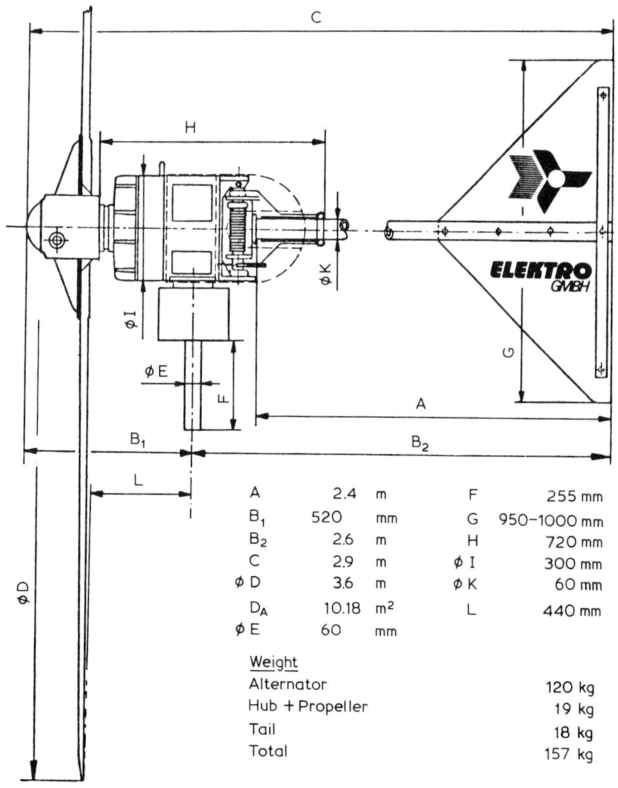

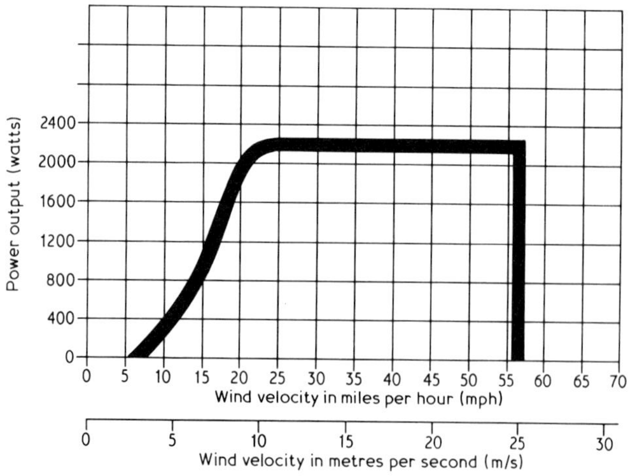

Elektro WV35G

PERFORMANCE AND OPERATIONAL DATA

Rated electrical power	3800–4000 watt
Rated output voltage	48/60/110 volt d.c.
Cut-in wind speed	2.5–3 m/s
Rated wind speed	10–12 m/s
Shut-down wind speed	20 m/s
Survival wind speed	60–70 m/s
Maximum shaft speed	450 rev/min

PHYSICAL DATA

Turbine

Type: Horizontal axis, upwind with tail vane
Dimensions
diameter 4.4 m

Blades

Construction: Wood, reinforced with aluminium
Number: 3

Control: Centrifugally operated blade pitch control; signal from auxiliary wind pressure switch used to turn turbine into the plane of the tail vane and wind

Generator: Permanent magnet a.c. generator (rectified)

Mechanical transmission: Direct coupling to turbine

System weight
net 235 kg
gross 330 kg

PRODUCTION

First production: 1950; earlier types varied in component detail and incorporated a d.c. generator

MISCELLANEOUS

Automatic shut-down in storm or with fully charged battery; type 274 unit needs manual restart, type 266 gives automatic restart with 6/12 hour time switch.

111–39a masthead (0.82 m) available for 3 or 4 leg towers, or mast tube only.

Several alternative tower types available, with mod. 137 assembly crane to suit.

Control panel, inverter and batteries also offered.

Annual maintenance comprises lubrication, overall check, paint touch-up (for severe sites and coastal locations twice a year is recommended).

One year warranty according to VSM (Swiss Association of Machinery Manufacturers).

VARIANTS

WV35D available, with rated output 3500 watt, 125–220 volt a.c. (35–70 Hz) using field rotor a.c. generator for electric heating.

3.9 m diameter version is offered for high wind speed or turbulent sites.

Elektro GmbH,
St Gallerstrasse 27,
Winterthur,
CH–8400, Switzerland
Telephone: (052) 22 34 34
Telex: 76299 Elektro

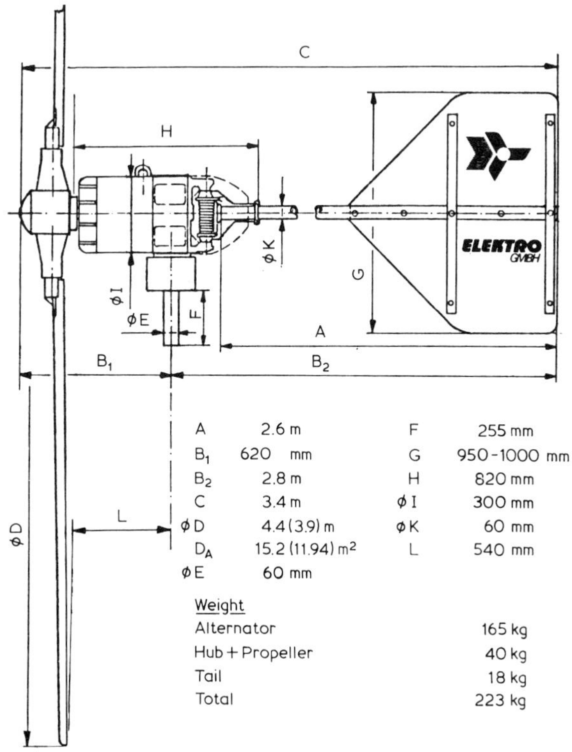

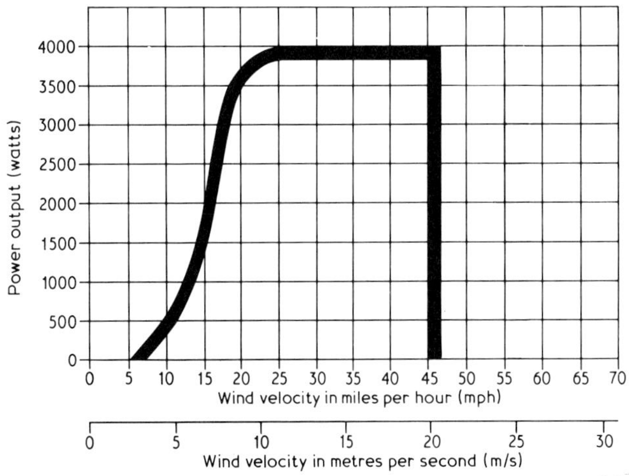

Elektro WVG 50G

PERFORMANCE AND OPERATIONAL DATA

Rated electrical power	5000–6000 watt
Rated output voltage	60/110/220 volt d.c.
Cut-in wind speed	2.5–3 m/s
Rated wind speed	10–12 m/s
Shut-down wind speed	20 m/s
Survival wind speed	60–70 m/s
Maximum shaft speed	250 rev/min

PHYSICAL DATA

Turbine

Type: Horizontal axis, upwind with tail vane
Dimensions
diameter 5.0 m

Blades

Construction: Wood, reinforced with aluminium
Number: 3

Control: Centrifugally operated blade pitch control; signal from auxiliary wind pressure switches used to turn turbine into the plane of the tail vane and wind

Generator: Permanent magnet a.c. generator (rectified)

Mechanical transmission: Speed increasing gearbox (internal teeth)

System weight
net 265 kg
gross 390 kg

PRODUCTION

First production: 1948; earlier types incorporated a d.c. generator

MISCELLANEOUS

Automatic shut-down in storm or with fully charged battery; type 274 unit needs manual restart, type 266 gives automatic restart with 6/12 hour time switch.
111–39a masthead (0.82 m) available for 3 or 4 leg towers, or mast tube only.
Several alternative tower types are available, with mod. 137 assembly crane to suit.
Control panel, inverter and batteries also offered.
Annual maintenance comprises lubrication, overall check, paint touch-up (for severe sites and coastal locations twice a year is recommended).
One year warranty according to VSM (Swiss Association of Machinery Manufacturers).

VARIANTS

WVG 50D available, with rated output 5000 watt, 125–220 volt a.c. (60–130 Hz) using field rotor a.c. generator with voltage regulator (type 300).

Elektro GmbH,
St Gallerstrasse 27,
Winterthur,
CH–8400, Switzerland
Telephone: (052) 22 34 34
Telex: 76299 Elektro

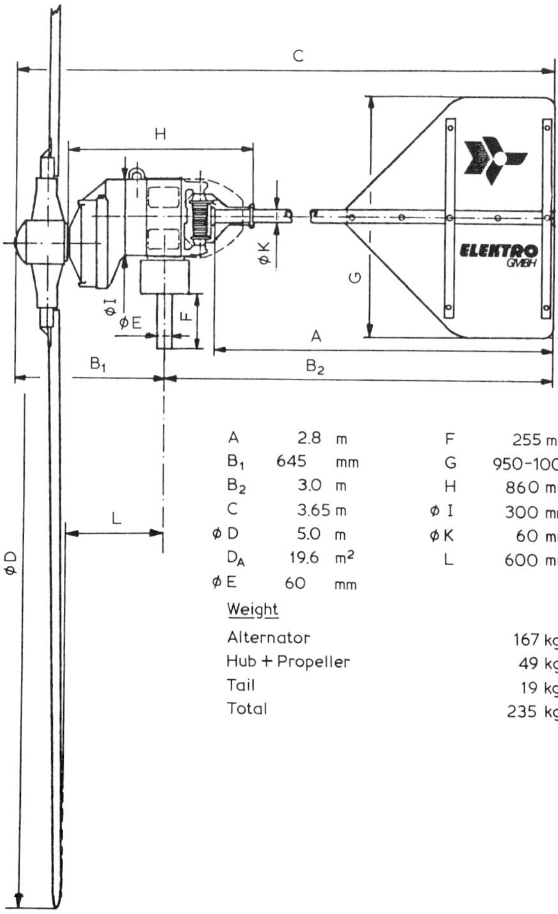

A	2.8 m	F	255 mm
B_1	645 mm	G	950–1000 mm
B_2	3.0 m	H	860 mm
C	3.65 m	Ø I	300 mm
Ø D	5.0 m	Ø K	60 mm
D_A	19.6 m^2	L	600 mm
Ø E	60 mm		

Weight

Alternator	167 kg
Hub + Propeller	49 kg
Tail	19 kg
Total	235 kg

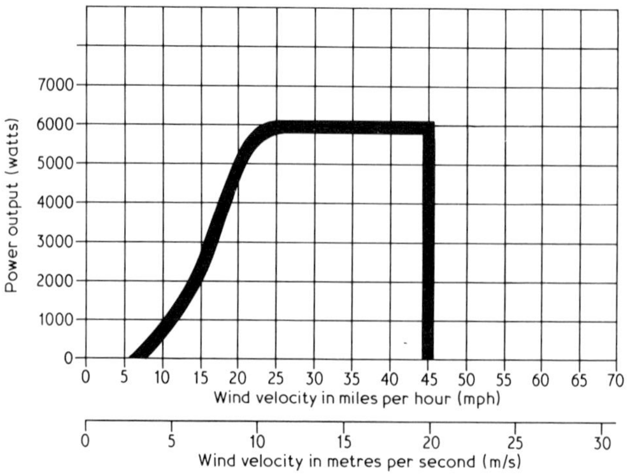

Elektro WVG 120G

PERFORMANCE AND OPERATIONAL DATA

Rated electrical power	8000–10 000 watt
Rated output voltage	110/220 volt d.c.
Cut-in wind speed	2.5–3 m/s
Rated wind speed	11–13 m/s
Shut-down wind speed	20 m/s
Survival wind speed	60–70 m/s
Maximum shaft speed	250 rev/min

PHYSICAL DATA

Turbine
Type: Horizontal axis, upwind with tail vane
Dimensions
diameter 6.0 m

Blades
Construction: Wood, reinforced with aluminium
Number: 3

Control: Centrifugally operated blade pitch control; signal from auxiliary wind pressure switches used to turn turbine into the plane of tail vane and wind

Generator: Permanent magnet a.c. generator (rectified)

Mechanical transmission: Speed increasing gearbox (internal teeth)

System weight
net 315 kg
gross 440 kg

PRODUCTION

First production: 1975

MISCELLANEOUS

Automatic shut-down in storm or with fully charged battery; type 274 unit needs manual restart, type 266 gives automatic restart with 6/12 hour time switch.

111–39a masthead (0.82 m) available for 3 or 4 leg towers, or mast tube only.

Several alternative tower types are available, with mod. 137 assembly crane to suit.

Control panel, inverter and batteries are offered.

Annual maintenance comprises lubrication, overall check, paint touch-up (for severe sites and coastal locations twice a year is recommended).

One year warranty according to VSM (Swiss Association of Machinery Manufacturers).

VARIANTS

WVG 120D available, with rated output 8000 watt, 125–220 volt a.c. (60–130 Hz) using field rotor a.c. generator for electric heating.

Alternative turbine diameter of 6.6 m is offered.

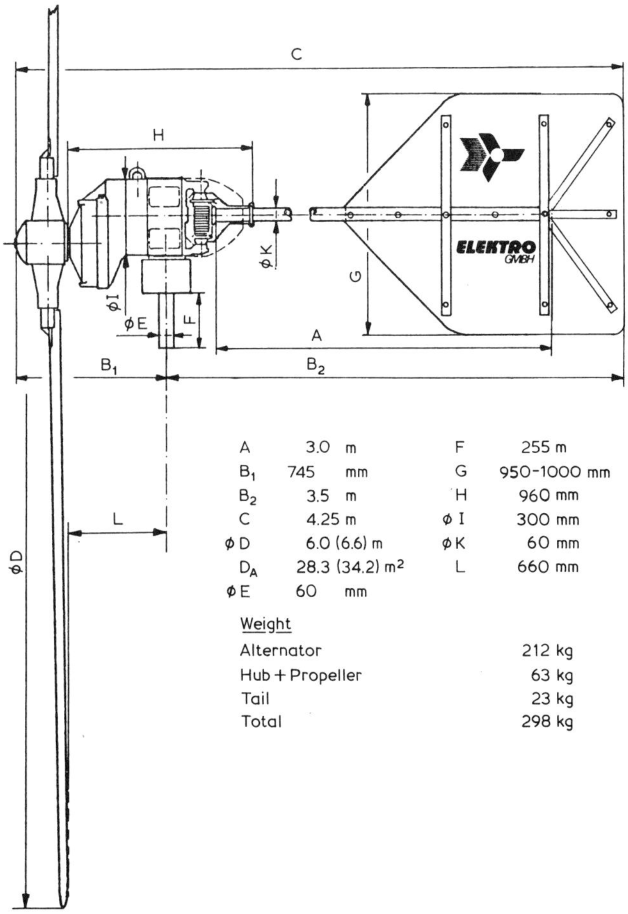

A	3.0 m	F	255 m
B_1	745 mm	G	950–1000 mm
B_2	3.5 m	H	960 mm
C	4.25 m	ϕ I	300 mm
ϕ D	6.0 (6.6) m	ϕ K	60 mm
D_A	28.3 (34.2) m²	L	660 mm
ϕ E	60 mm		

Weight	
Alternator	212 kg
Hub + Propeller	63 kg
Tail	23 kg
Total	298 kg

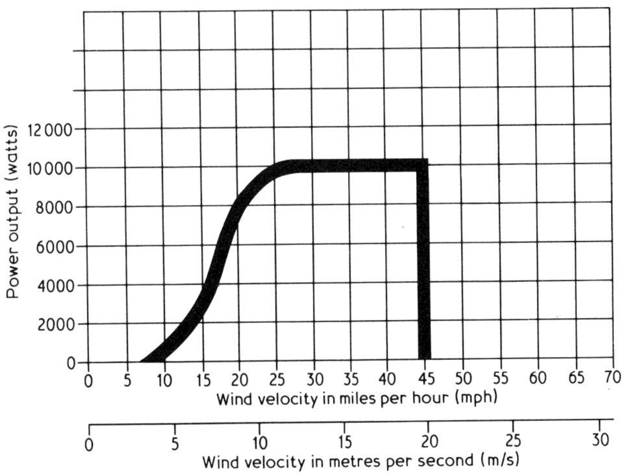

Elektro GmbH,
St Gallerstrasse 27,
Winterthur,
CH–8400, Switzerland
Telephone: (052) 22 34 34
Telex: 76299 Elektro

ENAG
Model 1

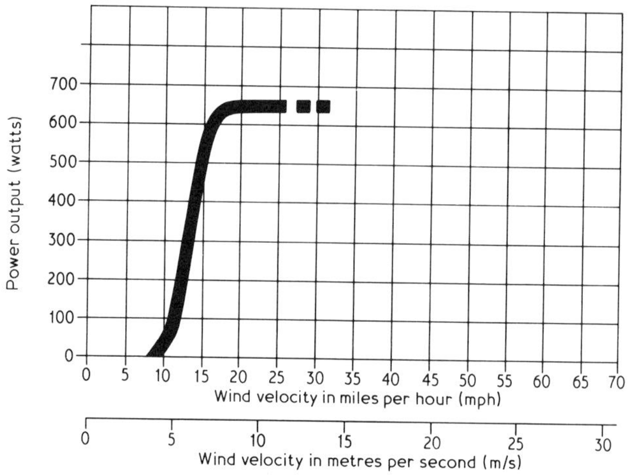

PERFORMANCE AND OPERATIONAL DATA

Rated electrical power	650 watt
Rated output voltage	24/30 volt d.c.
Cut-in wind speed	3.5 m/s
Rated wind speed	8 m/s
Maximum shaft speed	600 rev/min

PHYSICAL DATA

Turbine
Type: Horizontal axis, upwind with tail vane
Dimensions
diameter 2.55 m
Blades
Construction: Aluminium
Number: 2
Generator: d.c. generator
Mechanical transmission: Direct coupling to turbine

MISCELLANEOUS

Earlier variant was rated 400 watt.
Tubular tower head section of 1 m length offered.

ENAG S.A.,
Route de Pont-L'abbe,
29000–Quimper,
France
Telephone: 16(98) 95 44 25
16(98) 95 31 44

ENAG
Model 2

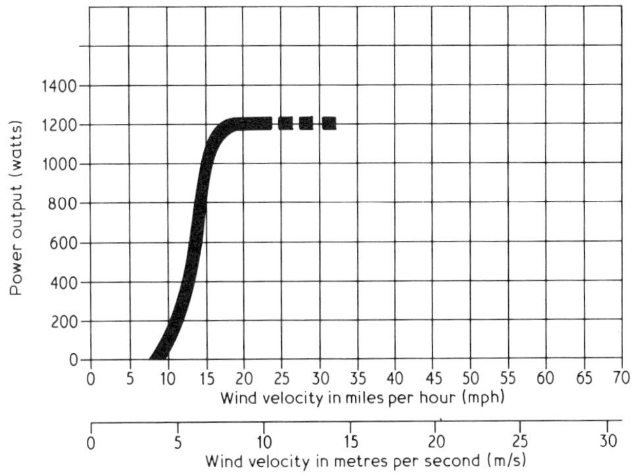

PERFORMANCE AND OPERATIONAL DATA

Rated electrical power	1200 watt
Rated output voltage	24/30 volt d.c.
Cut-in wind speed	3.5 m/s
Rated wind speed	8 m/s

PHYSICAL DATA

Turbine
Type: Horizontal axis, upwind with tail vane
Blades
Construction: Aluminium
Number: 2
Generator: d.c. generator
Mechanical transmission: Direct coupling to turbine

MISCELLANEOUS

Earlier variant had 3 blades.
Tubular tower head section of 1.1 m, 2.2 m, or 3 m length offered.

ENAG S.A.,
Route de Pont-L'abbe,
29000–Quimper,
France
Telephone: 16(98) 95 44 25
16(98) 95 31 44

ENAG
Model 3

PERFORMANCE AND OPERATIONAL DATA

Rated electrical power	2500 watt
Rated output voltage	110 volt d.c.
Cut-in wind speed	3.5 m/s
Rated wind speed	8 m/s

PHYSICAL DATA

Turbine
Type: Horizontal axis, upwind with tail vane
Blades
Construction: Aluminium
Number: 2
Generator: d.c. generator
Mechanical transmission: Direct coupling to turbine

MISCELLANEOUS

Earlier variant had 3 blades.
Tower head section of 1.5 m length offered.

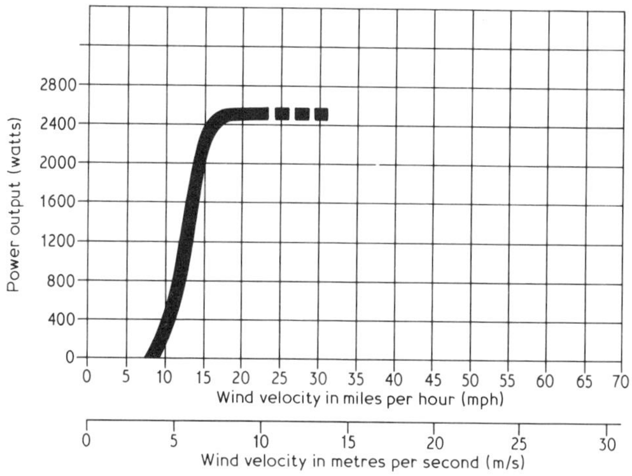

ENAG S.A.,
Route de Pont-L'abbe,
29000–Quimper,
France
Telephone: 16(98) 95 44 25
16(98) 95 31 44

Enertech 1800

PERFORMANCE AND OPERATIONAL DATA

Rated electrical power	1800 watt
Rated output voltage	115 volt a.c.
(operates in parallel with 60 Hz power network)	
Cut-in wind speed	3.6 m/s
Rated wind speed	10.7 m/s
Shut-down wind speed	17.9 m/s
Survival wind speed	53.7 m/s
Shaft speed at rated output	170 rev/min
Maximum shaft speed	178 rev/min

PHYSICAL DATA

Turbine

Type: Horizontal axis, downwind type, operating freely in yaw

Dimensions

diameter 3.96 m

Blades

Construction: Wooden, neoprene leading edge

Number: 3

Control: Fixed pitch turbine

Generator: Induction generator

Mechanical transmission: 11:1 ratio two-stage gearbox

System weight

net 120 kg

gross 170 kg

PRODUCTION

First production: Pre-production series 1978, full production 1979.

Production quantity to date: Over 450 (including preceding 1500 model) (May 1981).

MISCELLANEOUS

12.2 m, 21.4 m and 27.5 m guyed galvanized steel towers are offered, also guyed wood telephone pole or tubular pole (truss) tower.

Overspeed control by automatic brake.

Annual maintenance, oil change and system inspection.

Parts and workmanship guaranteed for one year.

REPORTS

Rocky Flats Performance Summary Sheet Enertech 1500, December 1980.

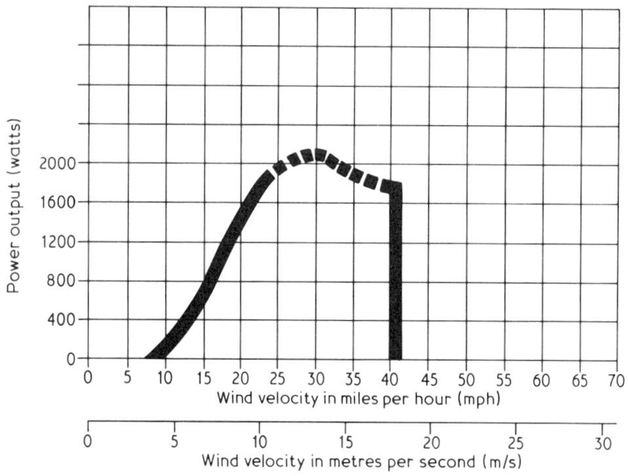

Enertech Corporation,
PO Box 420, Norwich
Vermont 05055
USA
Telephone: (802) 649 1145

Ideolec
50 x 7

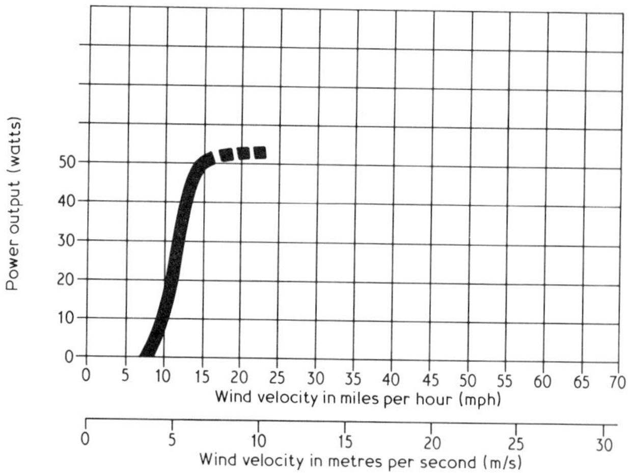

PERFORMANCE AND OPERATIONAL DATA

Rated electrical power	50 watt
Rated output voltage	12 volt d.c.
Cut-in wind speed	3 m/s
Rated wind speed	7 m/s
Shaft speed at rated output	600 rev/min

PHYSICAL DATA

Turbine
Type: Horizontal axis
Dimensions
diameter 1.2 m
Blades
Number: 2
Control: Fixed pitch turbine
Generator: a.c. generator
Mechanical transmission: Direct coupling to turbine
System weight
net 15 kg
gross 28 kg

MISCELLANEOUS

Supplied with 1.5 m (base-to-hub height) three-leg stub tower.

Eoliennes Humblot,
8, rue d'Alger, Coussey,
88300 Neufchateau,
France
Telephone: (29) 94 09 09

Ideolec
50 x 8

PERFORMANCE AND OPERATIONAL DATA

Rated electrical power	50 watt
Rated output voltage	24 volt d.c.
Cut-in wind speed	3 m/s
Rated wind speed	8 m/s

PHYSICAL DATA

Turbine
Type: Horizontal axis
Dimensions
diameter 0.85 m
Blades
Number: 2
Control: Fixed pitch turbine
Generator: d.c. generator
Mechanical transmission: Speed increasing transmission
System weight
net 17 kg
gross 30 kg

MISCELLANEOUS

Supplied with 500 mm diameter tubular tower.

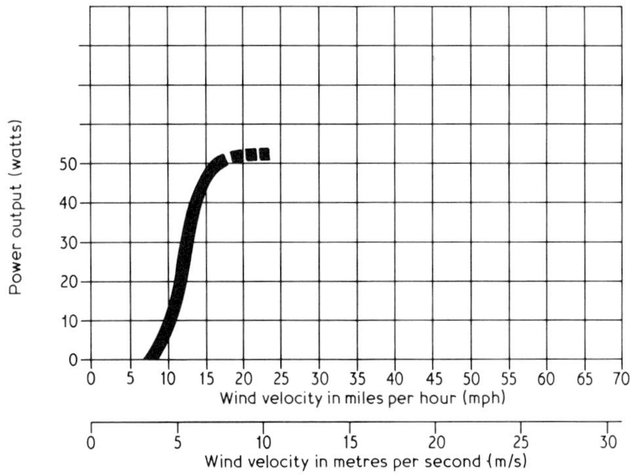

Eoliennes Humblot,
8, rue d'Alger, Coussey,
88300 Neufchateau,
France
Telephone: (29) 94 09 09

Ideolec
100 x 4

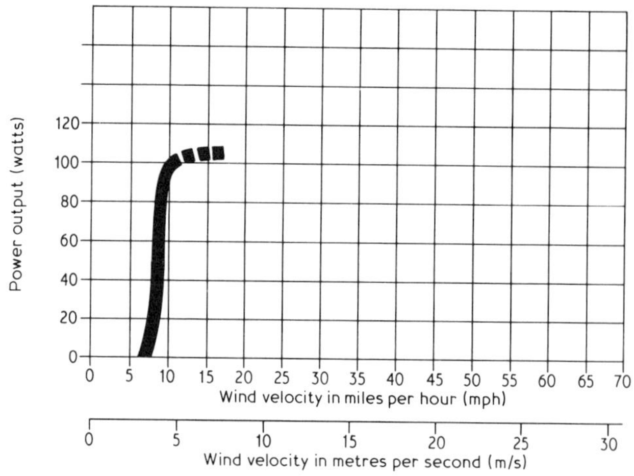

PERFORMANCE AND OPERATIONAL DATA

Rated electrical power	100 watt
Rated output voltage	27 volt d.c.
Cut-in wind speed	3 m/s
Rated wind speed	4 m/s

PHYSICAL DATA

Turbine
Type: Horizontal axis, upwind with single 35 mm diameter fan tail
Dimensions
diameter 3.10 m
Blades
Number: 3
Control: Fixed pitch turbine
Generator: d.c. generator
Mechanical transmission: 7:1 ratio speed increasing (ratio adjustable)
System weight
net 141 kg

MISCELLANEOUS

Four-leg lattice towers provided, either 2.70 m stub tower or 5.80 m complete tower.

Eoliennes Humblot,
8, rue d'Alger, Coussey,
88300 Neufchateau,
France
Telephone: (29) 94 09 09

Ideolec
100 x 5

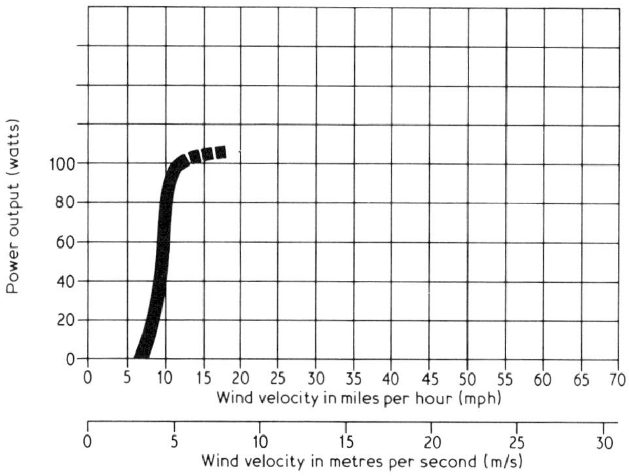

PERFORMANCE AND OPERATIONAL DATA

Rated electrical power	100 watt
Rated output voltage	27 volt d.c.
Cut-in wind speed	3 m/s
Rated wind speed	5 m/s

PHYSICAL DATA

Turbine
Type: Horizontal axis, upwind with single 35 mm diameter fan tail
Dimensions
diameter 2.50 m
Blades
Number: 3
Control: Fixed pitch turbine
Generator: d.c. generator
Mechanical transmission: Direct coupling to turbine
System weight
net 98 kg

MISCELLANEOUS

Four-leg lattice towers provided, either 2.70 m stub tower or 5.80 m complete tower.

Eoliennes Humblot,
8, rue d'Alger, Coussey,
88300 Neufchateau,
France
Telephone: (29) 94 09 09

Ideolec 100 x 6

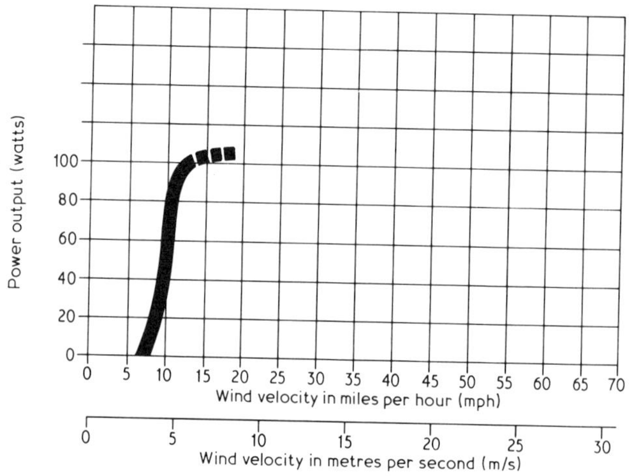

PERFORMANCE AND OPERATIONAL DATA

Rated electrical power	100 watt
Rated output voltage	12/24 volt d.c.
Cut-in wind speed	3 m/s
Rated wind speed	6 m/s

PHYSICAL DATA

Turbine
Type: Horizontal axis, upwind with single 35 mm diameter fan tail
Dimensions
diameter 2.50 m
Blades
Number: 3
Control: Fixed pitch turbine
Generator: d.c. generator
Mechanical transmission: 7:1 ratio speed increasing (ratio adjustable)
System weight
net 95 kg

MISCELLANEOUS

Four-leg lattice towers provided, either 2.70 m stub tower or 5.80 m complete tower.

Eoliennes Humblot,
8, rue d'Alger, Coussey,
88300 Neufchateau,
France
Telephone: (29) 94 09 09

Ideolec 300 x 7

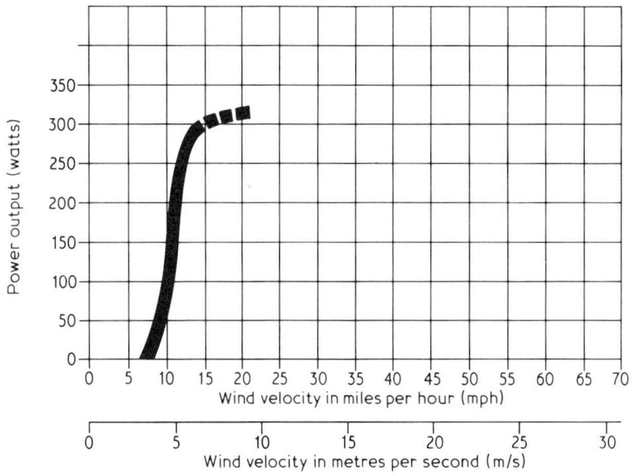

PERFORMANCE AND OPERATIONAL DATA

Rated electrical power	300 watt
Rated output voltage	12 volt d.c.
Cut-in wind speed	3 m/s
Rated wind speed	7 m/s

PHYSICAL DATA

Turbine
Type: Horizontal axis, upwind with single 30 mm diameter fan tail
Dimensions
diameter 2.50 m
Blades
Number: 3
Control: Fixed pitch turbine
Generator: a.c. generator
Mechanical transmission: Direct coupling to turbine
System weight
net 60 kg

MISCELLANEOUS

Four-leg lattice towers provided, either 2.70 m stub tower or 5.80 m complete tower.

Eoliennes Humblot,
8, rue d'Alger, Coussey,
88300 Neufchateau,
France
Telephone: (29) 94 09 09

Ideolec
500 x 7

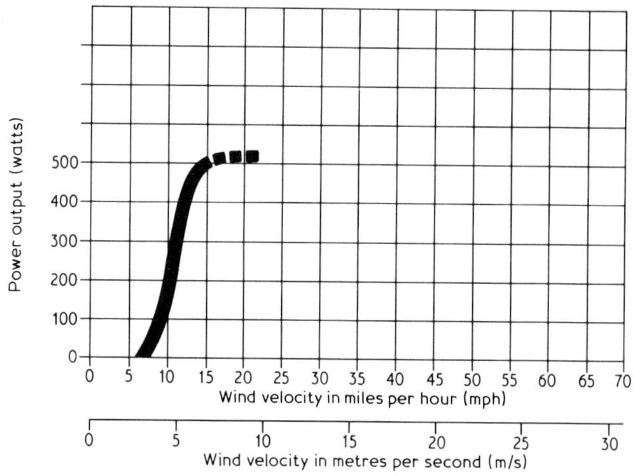

PERFORMANCE AND OPERATIONAL DATA

Rated electrical power	500 watt
Rated output voltage	24 volt d.c.
Cut-in wind speed	3 m/s
Rated wind speed	7 m/s

PHYSICAL DATA

Turbine
Type: Horizontal axis, upwind with single 35 mm diameter fan tail
Dimensions
diameter 3.10 m
Blades
Number: 4
Control: Fixed pitch turbine
Generator: d.c. generator
Mechanical transmission: 7:1 ratio speed increasing (ratio adjustable)
System weight
net 137 kg

MISCELLANEOUS

Four-leg lattice towers provided, either 2.70 m stub tower or 5.80 m complete tower.

Eoliennes Humblot,
8, rue d'Alger, Coussey,
88300 Neufchateau,
France
Telephone: (29) 94 09 09

Ideolec 1000 x 6

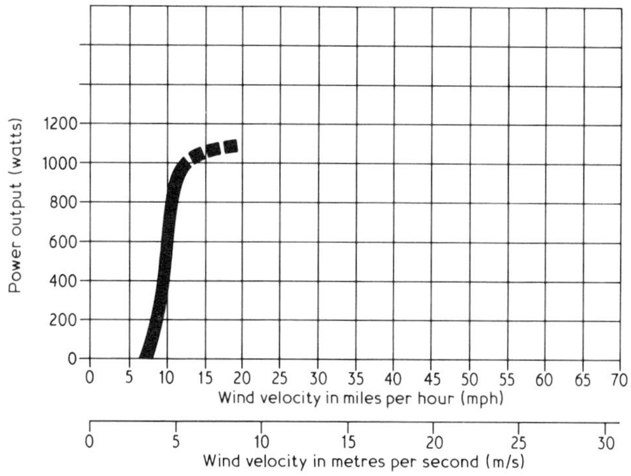

PERFORMANCE AND OPERATIONAL DATA

Rated electrical power	1000 watt
Rated output voltage	24 volt d.c.
Cut-in wind speed	3 m/s
Rated wind speed	6 m/s

PHYSICAL DATA

Turbine

Type: Horizontal axis, upwind with twin 50 mm diameter fan tails

Dimensions

diameter 4.60 m

Blades

Number: 3

Control: Variable pitch turbine

Generator: d.c. generator

Mechanical transmission: 10:1 ratio speed increasing (ratio adjustable)

System weight

net 416 kg

MISCELLANEOUS

Four-leg lattice towers provided, either 5.30 m stub tower or 9.30 m or 13.30 m complete towers.

Eoliennes Humblot,
8, rue d'Alger, Coussey,
88300 Neufchateau,
France
Telephone: (29) 94 09 09

Ideolec 1000 x 7

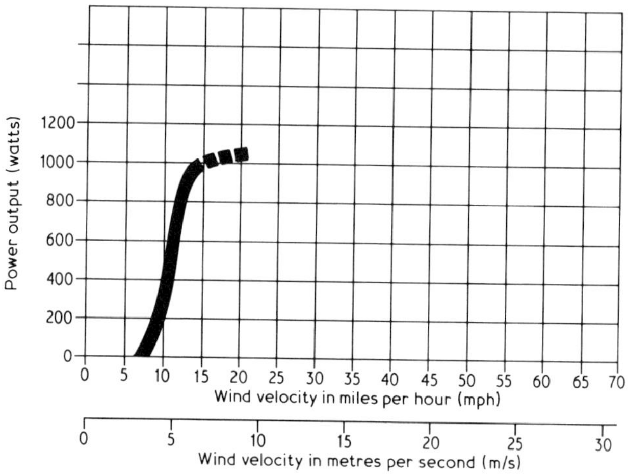

PERFORMANCE AND OPERATIONAL DATA

Rated electrical power	1000 watt
Rated output voltage	24 volt d.c.
Cut-in wind speed	3 m/s
Rated wind speed	7 m/s

PHYSICAL DATA

Turbine
Type: Horizontal axis, upwind with twin 50 mm diameter fan tails
Dimensions
diameter 4.20 m
Blades
Number: 2
Control: Variable pitch turbine
Generator: d.c. generator
Mechanical transmission: 10:1 ratio speed increasing (ratio adjustable)
System weight
net 393 kg

MISCELLANEOUS

Four-leg lattice towers provided, either 5.30 m stub tower or 9.30 m or 13.30 m complete towers.

Eoliennes Humblot,
8, rue d'Alger, Coussey,
88300 Neufchateau,
France
Telephone: (29) 94 09 09

Ideolec
2000 x 7

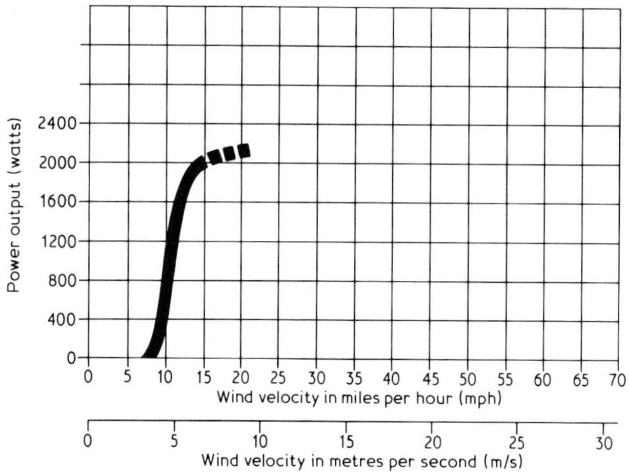

PERFORMANCE AND OPERATIONAL DATA

Rated electrical power	2000 watt
Rated output voltage	220 volt a.c.
Cut-in wind speed	3 m/s
Rated wind speed	7 m/s

PHYSICAL DATA

Turbine

Type: Horizontal axis, upwind with twin 60 mm diameter fan tails

Dimensions

diameter 5.20 m

Blades

Number: 3

Control: Variable pitch turbine

Generator: a.c. generator

Mechanical transmission: 10:1 ratio speed increasing (ratio adjustable)

System weight

net 432 kg

MISCELLANEOUS

Four-leg lattice towers provided, either 5.30 m stub tower or 9.30 m or 13.30 m complete towers.

Eoliennes Humblot,
8, rue d'Alger, Coussey,
88300 Neufchateau,
France
Telephone: (29) 94 09 09

Ideolec
3000 x 6

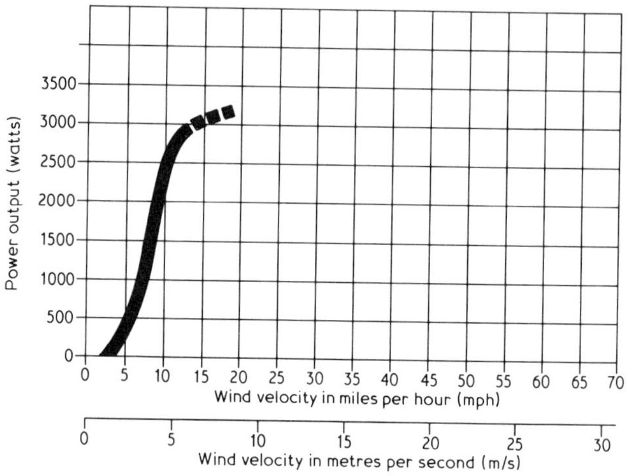

PERFORMANCE AND OPERATIONAL DATA

Rated electrical power	3000 watt
Rated output voltage	48 volt d.c.
Cut-in wind speed	1 m/s
Rated wind speed	6 m/s

PHYSICAL DATA

Turbine
Type: Horizontal axis type, upwind with twin 60 mm diameter fan tail
Dimensions
diameter 8.00 m
Blades
Number: 3
Control: Variable pitch turbine
Generator: d.c. generator
Mechanical transmission: 80:1 ratio speed increasing (ratio adjustable)
System weight
net 766 kg

MISCELLANEOUS

9.30 m or 13.30 m four-leg lattice towers provided.

Eoliennes Humblot,
8, rue d'Alger, Coussey,
88300 Neufchateau,
France
Telephone: (29) 94 09 09

Ideolec 3000 x 8

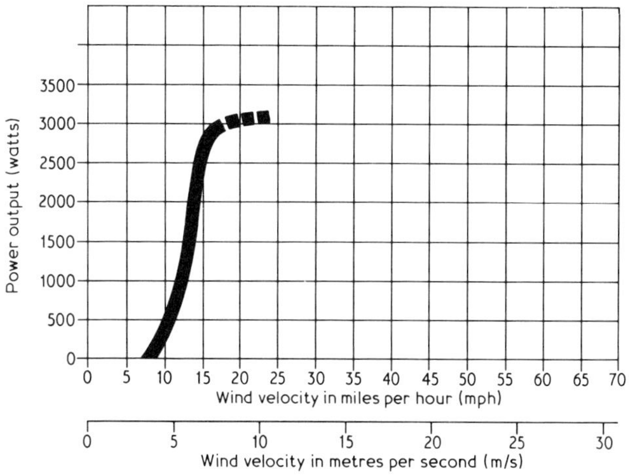

PERFORMANCE AND OPERATIONAL DATA

Rated electrical power	3000 watt
Rated output voltage	48 volt d.c.
Cut-in wind speed	3 m/s
Rated wind speed	8 m/s

PHYSICAL DATA

Turbine
Type: Horizontal axis, upwind with twin 60 mm diameter fan tails
Dimensions
diameter 5.20 m
Blades
Number: 3
Control: Variable pitch turbine
Generator: d.c. generator
Mechanical transmission: 80:1 ratio speed increasing (ratio adjustable)
System weight
net 605 kg

MISCELLANEOUS

Four-leg lattice towers provided, either 5.30 m stub tower or 9.30 m or 13.30 m complete towers.

Eoliennes Humblot,
8, rue d'Alger, Coussey,
88300 Neufchateau,
France
Telephone: (29) 94 09 09

Ideolec
5000 x 7

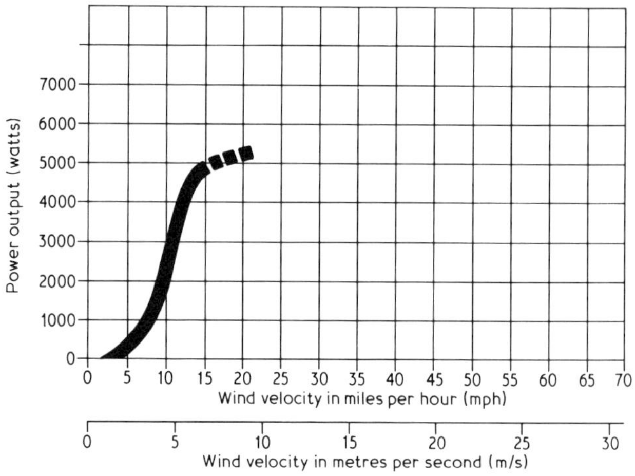

PERFORMANCE AND OPERATIONAL DATA

Rated electrical power	5000 watt
Rated output voltage	360 volt d.c.
Cut-in wind speed	1 m/s
Rated wind speed	7 m/s

PHYSICAL DATA

Turbine
Type: Horizontal axis, upwind with twin 60 mm diameter fan tails
Dimensions
diameter 8.00 m
Blades
Number: 3
Control: Variable pitch turbine
Generator: d.c. generator
Mechanical transmission: 80:1 ratio speed increasing (ratio adjustable)
System weight
net 836 kg

MISCELLANEOUS

9.30 m or 13.30 m four-leg lattice towers provided.

Eoliennes Humblot,
8, rue d'Alger, Coussey,
88300 Neufchateau,
France
Telephone: (29) 94 09 09

Fuji
5 KW

PERFORMANCE AND OPERATIONAL DATA

Rated electrical power	750 watt
Rated output voltage	29 volt d.c.
Cut-in wind speed	3.0 m/s
Rated wind speed	6.0 m/s
Shut-down wind speed	60 m/s
Survival wind speed	60 m/s
Shaft speed at rated output	120 rev/min

PHYSICAL DATA

Turbine
Type: Horizontal axis, upwind with tail vane
Dimensions
diameter 5.2 m
Blades
Construction: Steel plate
Number: 3
Control: Variable pitch turbine
Generator: 3 phase induction generator
Mechanical transmission: 5:1 ratio speed increasing

PRODUCTION

First production: 1978
Production quantity to date: 5

MISCELLANEOUS

Tubular tower pivoted on low tripod.

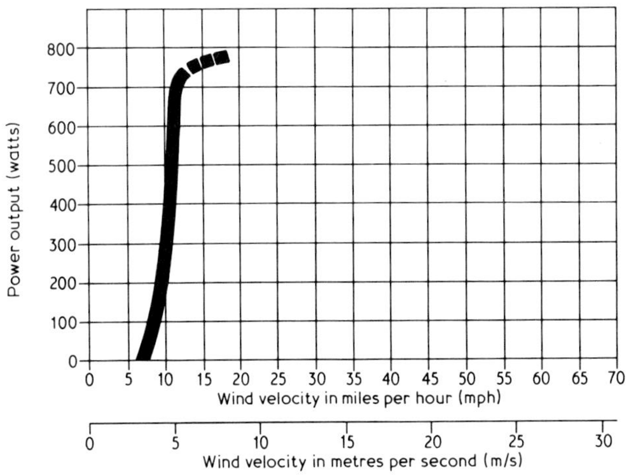

Fuji Electric Co. Ltd,
1–1 Tanabe-Shinden,
Kawasaki-ku, Kawasaki-shi,
Japan

Guanaco Aero-Gen

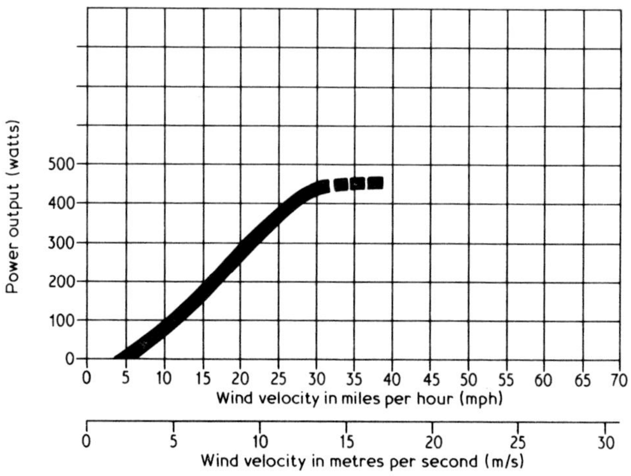

PERFORMANCE AND OPERATIONAL DATA

Rated electrical power	440 watt
Rated output voltage	12 volt d.c.
Cut-in wind speed	1.8 m/s
Rated wind speed	13.9 m/s
Shut-down wind speed	none
Survival wind speed	55.6 m/s
Maximum shaft speed	600 rev/min

PHYSICAL DATA

Turbine
Type: Horizontal axis, upwind with tail vane
Dimensions
diameter 1.90 m
Blades
Construction: Varnished wood, selected and seasoned
Number: 2
Control: Fixed pitch blades, with centrifugally operated regulator
Generator: d.c. generator
Mechanical transmission: Direct coupling to turbine
System weight
net 40 kg

PRODUCTION

First production: 1976
Production quantity to date: About 1500 (August 1979)

MISCELLANEOUS

This company is a leading manufacturer of wind pumps, of which it is claimed there are more than 500 000 in Argentina.
No maintenance specified.

Guanaco Internacional SRL,
Maipu 325, 1006 Buenos Aires,
Argentina
Telephone: 45–9686

Selectromarine 5 W

PERFORMANCE AND OPERATIONAL DATA

Rated electrical power	5 watt
Rated output voltage	12/24 volt d.c.
Cut-in wind speed	3.5–4.7 m/s
Rated wind speed	10.8–15.3 m/s
Shut-down wind speed	none
Survival wind speed	over 50 m/s
Maximum shaft speed	2000 rev/min

PHYSICAL DATA

Turbine

Type: Horizontal axis, multiblade, upwind with tail vane

Dimensions

diameter 0.43 m

Blades

Construction: Pressed and twisted from aluminium alloy

Number: 14

Control: Fixed pitch blades

Generator: Permanent magnet a.c. generator (rectified)

Mechanical transmission: Direct coupling to turbine

System weight

net 10 kg

PRODUCTION

First production: 1972

MISCELLANEOUS

Annual maintenance comprises cleaning the slip ring with a general solvent; five yearly inspection of brushes.

VARIANT

Two units are offered vertically in line on a common mounting tube as a 10 watt rated system. a.c. generator outputs are paralleled after rectification.

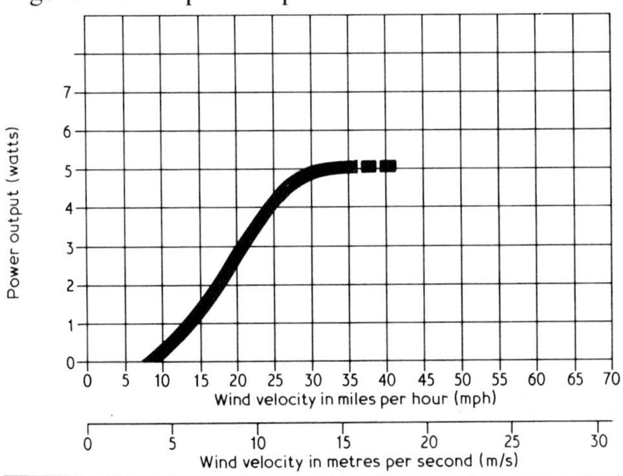

Howe International,
Nuffield Industrial Estate,
Poole, Dorset, England
Telephone: (0202) 685151
Telex: 41495
Cable: Ralphowe Poole

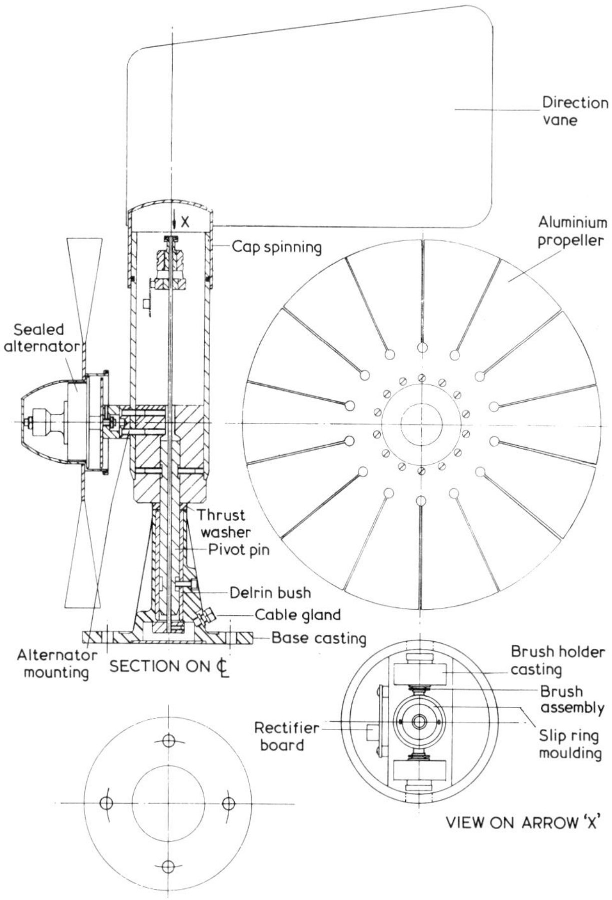

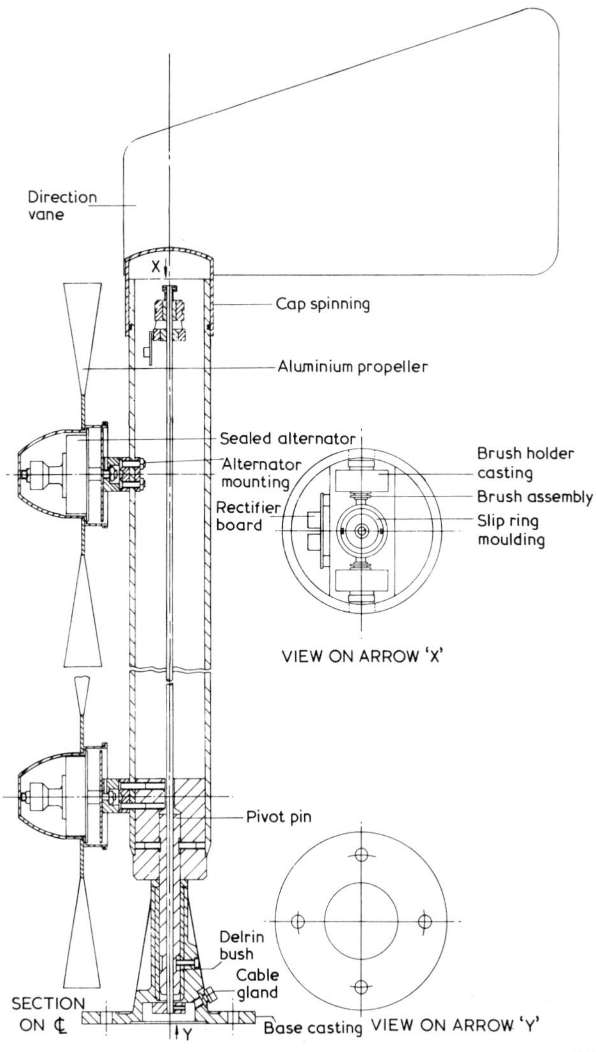

Selectromarine 50 W

PERFORMANCE AND OPERATIONAL DATA

Rated electrical power	50 watt
Rated output voltage	12/24 volt d.c.
Cut-in wind speed	4.4 m/s
Rated wind speed	17.5–17.8 m/s
Shut-down wind speed	none
Survival wind speed	over 50 m/s
Maximum shaft speed	2500 rev/min

PHYSICAL DATA

Turbine

Type: Horizontal axis, multiblade, upwind with tail vane

Dimensions

diameter 0.67 m

Blades

Construction: Polypropylene mounted on aluminium alloy hub

Number: 14

Control: Fixed pitch blades

Generator: Permanent magnet a.c. generator (rectified)

Mechanical transmission: Direct coupling to turbine

System weight

net 15.9 kg

PRODUCTION

First production: 1972

MISCELLANEOUS

Annual maintenance comprises cleaning the slip ring with a general solvent; five yearly inspection of brushes.

VARIANTS

Two units are offered vertically in line on a common mounting tube as a 100 watt rated system. a.c. generator outputs are paralleled after rectification.

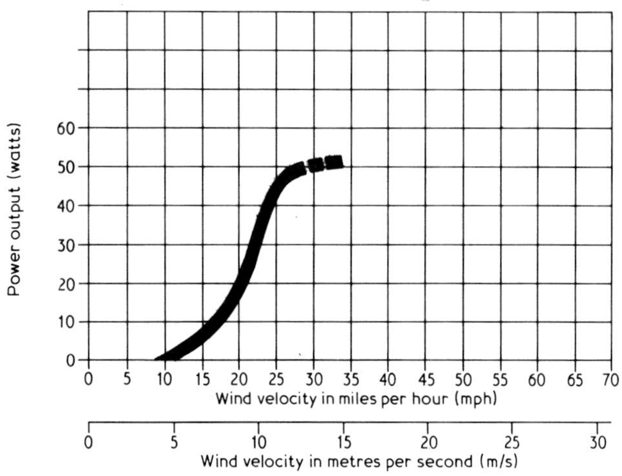

Howe International,
Nuffield Industrial Estate,
Poole, Dorset, England
Telephone: (0202) 685151
Telex: 41495
Cable: Ralphowe Poole

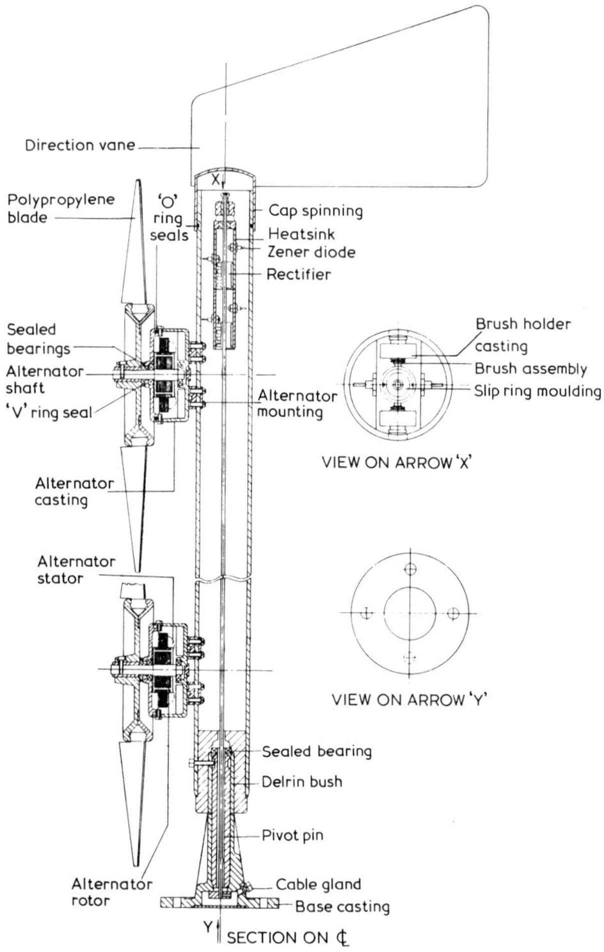

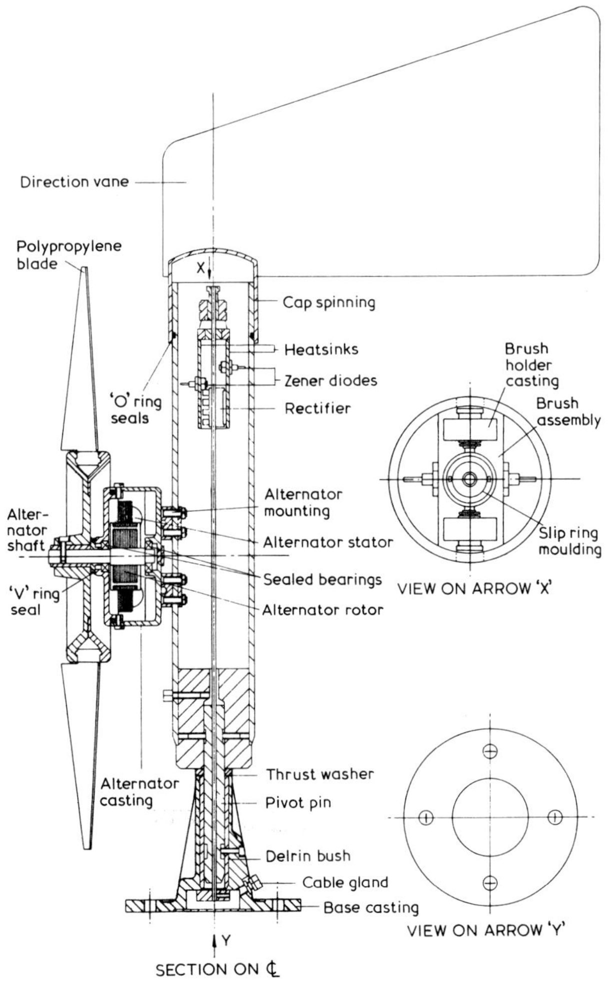

Fritz Hüllmann
HS 05S

PERFORMANCE AND OPERATIONAL DATA

Rated electrical power	5000 watt
Rated output voltage	220/380 volt
Cut-in wind speed	3 m/s
Rated wind speed	10 m/s
Shut-down wind speed	none
Survival wind speed	32.6 m/s
Shaft speed at rated output	150 rev/min

PHYSICAL DATA

Turbine
Type: Horizontal axis, downwind
Dimensions
diameter 5 m
Blades
Construction: GFK
Number: 3
Control: Variable pitch turbine
Generator: Brushless self-excited a.c. generator
Mechanical transmission: 10:1 ratio speed increasing gearbox
System weight: 1200 kg

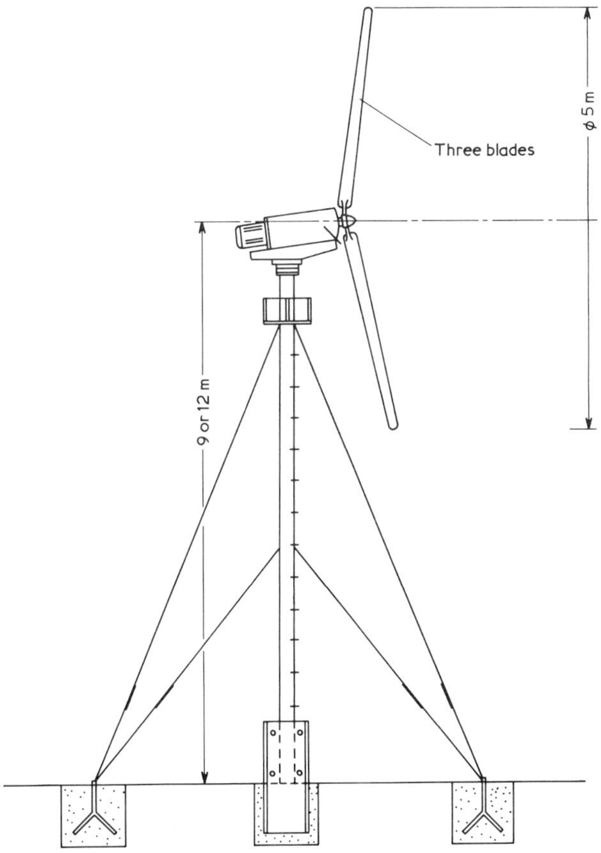

PRODUCTION

First production: 1978
Production quantity to date: 2 (April 1980)

MISCELLANEOUS

12 month warranty.
Handbook available.
9 m or 12 m high guyed tubular tower with maintenance platform.

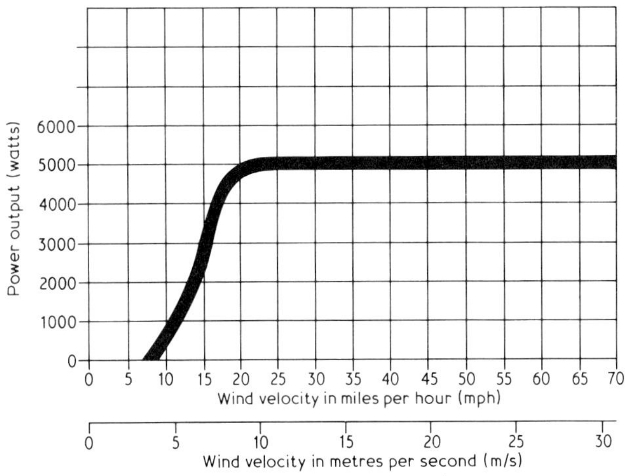

Fritz Hüllmann-Windkraftanlagen,
In de Hörn 1, D–2082 Tornesch,
West Germany
Telephone: 04122/52350

Fritz Hüllmann
HS 10S

PERFORMANCE AND OPERATIONAL DATA

Rated electrical power	10 000 watt
Rated output voltage	220/380 volt
Cut-in wind speed	3 m/s
Rated wind speed	10 m/s
Shut-down wind speed	none
Survival wind speed	32.6 m/s
Shaft speed at rated output	100 rev/min

PHYSICAL DATA

Turbine
Type: Horizontal axis, downwind
Dimensions
diameter 9 m
Blades
Construction: GFK
Number: 3
Control: Variable pitch turbine
Generator: Brushless self-excited a.c. generator
Mechanical transmission: Speed increasing gearbox
System weight: 2000 kg

PRODUCTION

First production: 1978
Production quantity to date: 2 (April 1980)

MISCELLANEOUS

12 month warranty.
Handbook available.
12 m or 15 m high guyed tubular tower with maintenance platform.

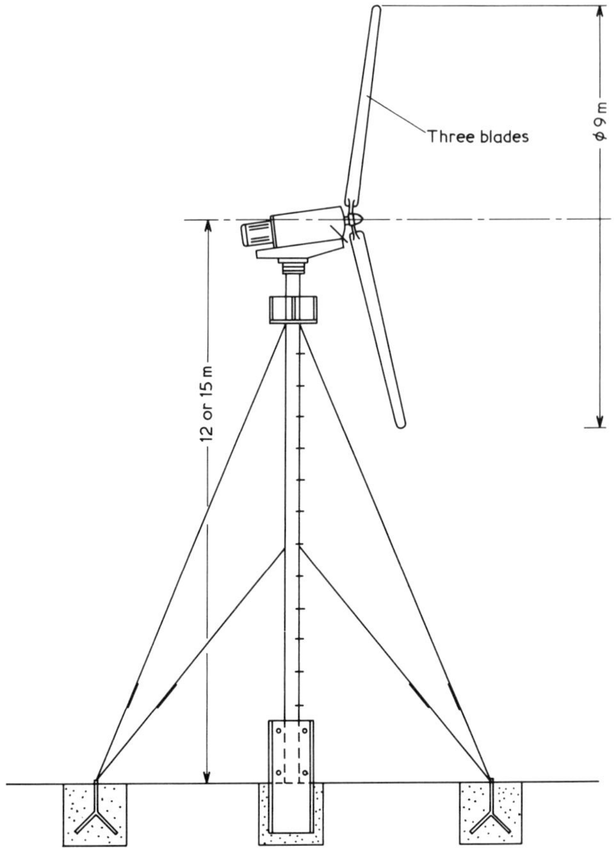

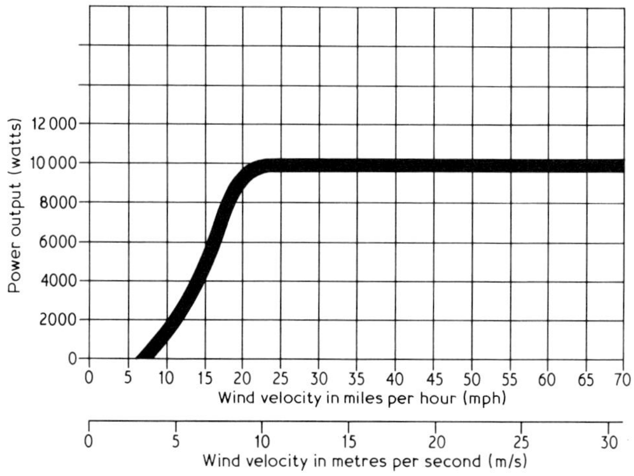

Fritz Hüllmann-Windkraftanlagen,
In de Hörn 1, D–2082 Tornesch,
West Germany
Telephone: 04122/52350

Fritz Hüllmann
HS 20S

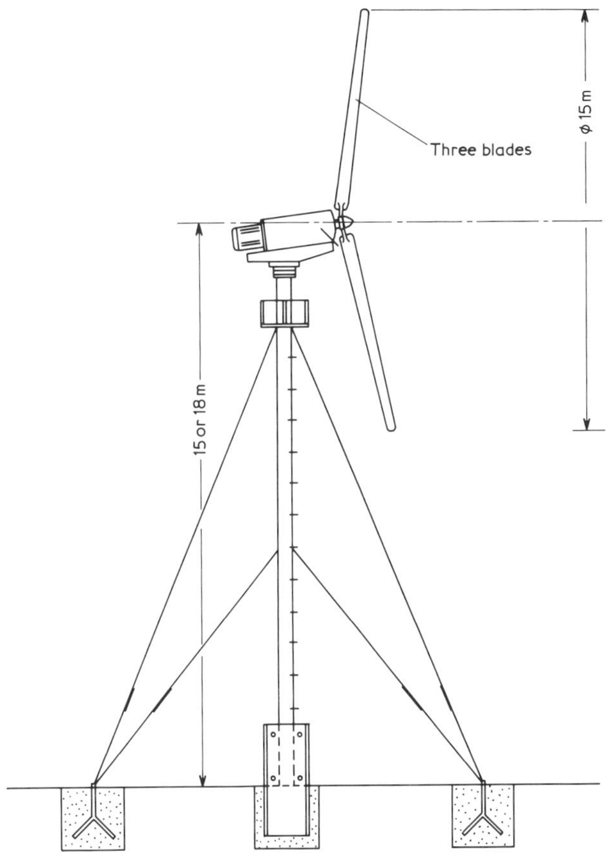

PERFORMANCE AND OPERATIONAL DATA

Rated electrical power	20000 watt
Rated output voltage	220/380 volt
Cut-in wind speed	3 m/s
Rated wind speed	10 m/s
Shut-down wind speed	none
Survival wind speed	32.6 m/s
Shaft speed at rated output	60 rev/min

PHYSICAL DATA

Turbine
Type: Horizontal axis, downwind
Dimensions
diameter 15 m
Blades
Construction: GFK
Number: 3
Control: Variable pitch turbine
Generator: Brushless self-excited a.c. generator
Mechanical transmission: Speed increasing gearbox
System weight: 3000 kg

PRODUCTION

First production: 1980
Production quantity to date: 2 (April 1980)

MISCELLANEOUS

12 month warranty.
Handbook available.
15 m or 18 m high guyed tubular tower with maintenance platform.

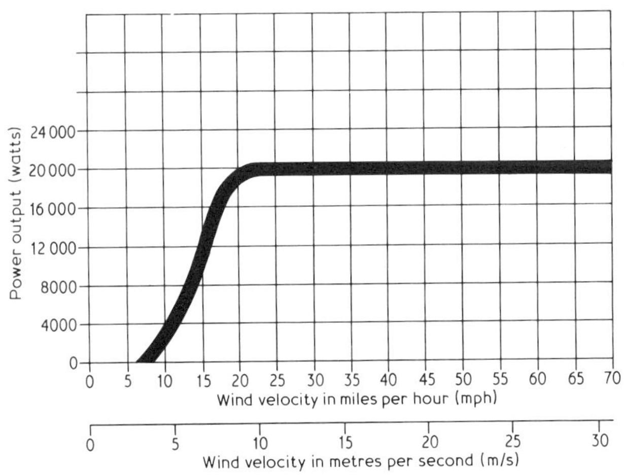

Fritz Hüllmann-Windkraftanlagen,
In de Hörn 1, D–2082 Tornesch,
West Germany
Telephone: 04122/52350

Fritz Hüllmann HM 05S

PERFORMANCE AND OPERATIONAL DATA

Rated electrical power	5000 watt
Rated output voltage	220/380 volt
Cut-in wind speed	2 m/s
Rated wind speed	10 m/s
Shut-down wind speed	none
Survival wind speed	32.6 m/s
Shaft speed at rated output	60 rev/min

PHYSICAL DATA

Turbine
Type: Upwind with tail vane
Dimensions
diameter 5 m
Blades
Construction: Aluminium
Number: 8
Control: Fixed pitch turbine
Generator: Brushless self-excited a.c. generator
Mechanical transmission: Speed increasing gearbox
System weight: 1500 kg

PRODUCTION

First production: 1978
Production quantity to date: 3 (April 1980)

MISCELLANEOUS

12 month warranty.
Handbook available.
9 m or 12 m high lattice steel tower.

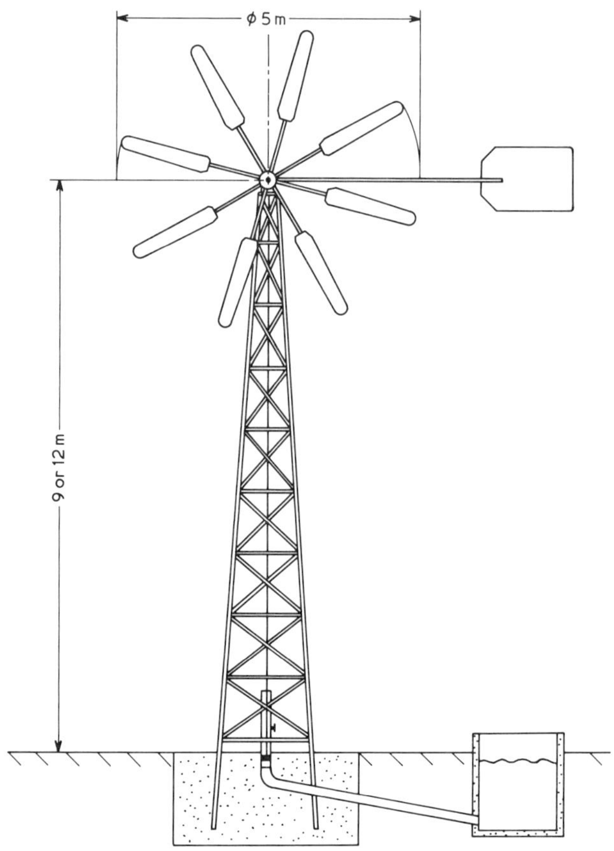

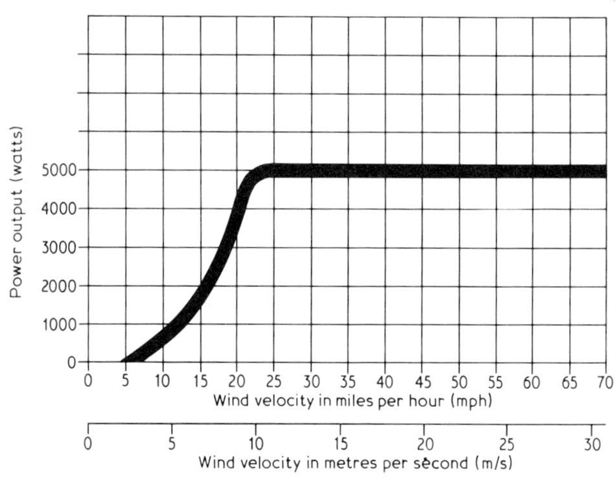

Fritz Hüllmann-Windkraftanlagen,
In de Hörn 1, D–2082 Tornesch,
West Germany
Telephone: 04122/52350

Fritz Hüllmann HM 10S

PERFORMANCE AND OPERATIONAL DATA

Rated electrical power	10 000 watt
Rated output voltage	220/380 volt
Cut-in wind speed	2 m/s
Rated wind speed	10 m/s
Shut-down wind speed	none
Survival wind speed	32.6 m/s
Shaft speed at rated output	40 rev/min

PHYSICAL DATA

Turbine
Type: Upwind with tail vane
Dimensions
diameter 9 m
Blades
Construction: Aluminium
Number: 10
Control: Fixed pitch turbine
Generator: Brushless self-excited a.c. generator
Mechanical transmission: Speed increasing gearbox
System weight: 2500 kg

PRODUCTION

First production: 1978
Production quantity to date: 3 (April 1980)

MISCELLANEOUS

12 month warranty.
Handbook available.
12 m or 15 m high lattice steel tower.

Fritz Hüllmann-Windkraftanlagen,
In de Hörn 1, D–2082 Tornesch,
West Germany
Telephone: 04122/52350

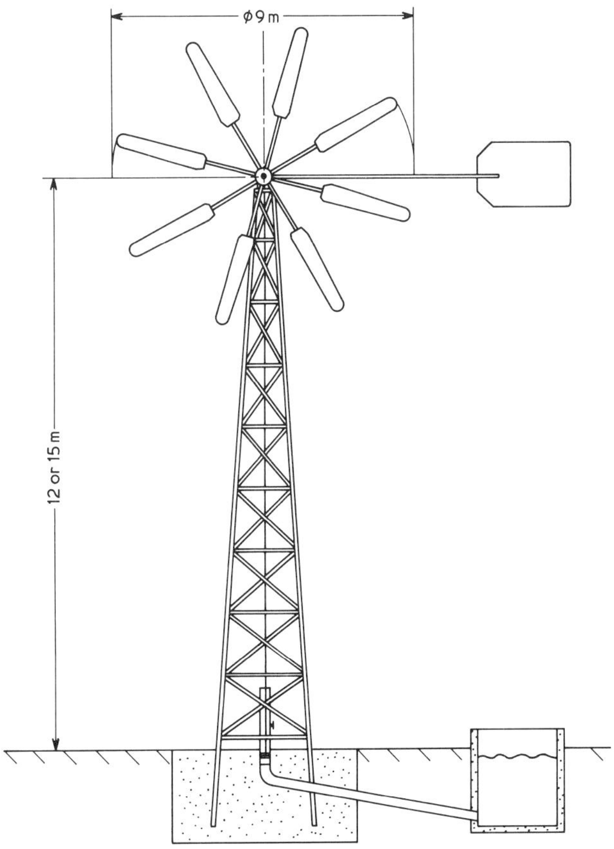

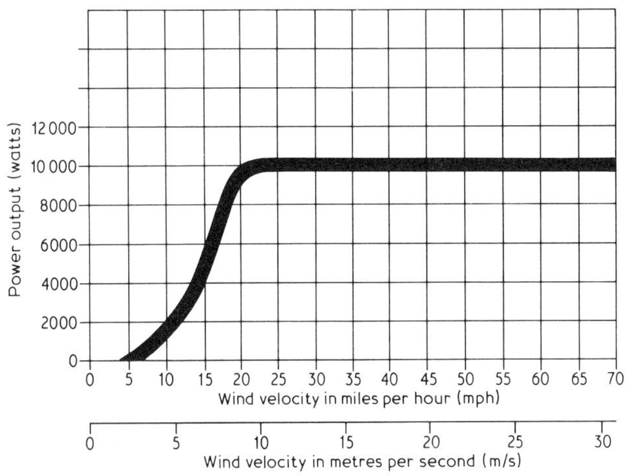

Independent Energy Systems Sky Hawk IV

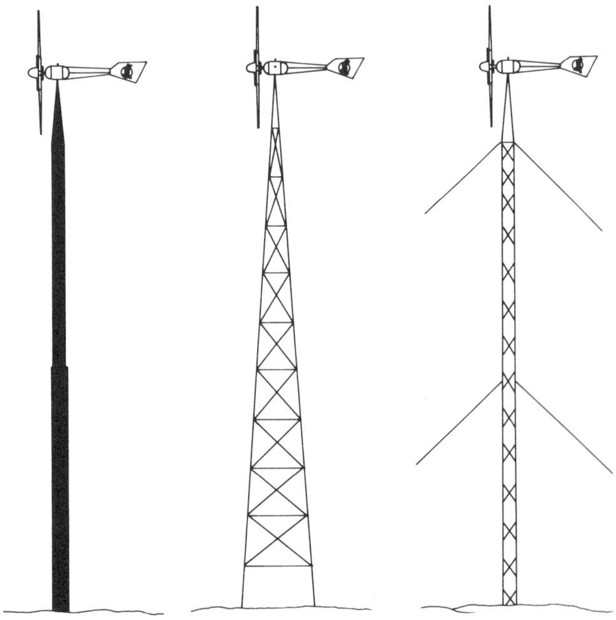

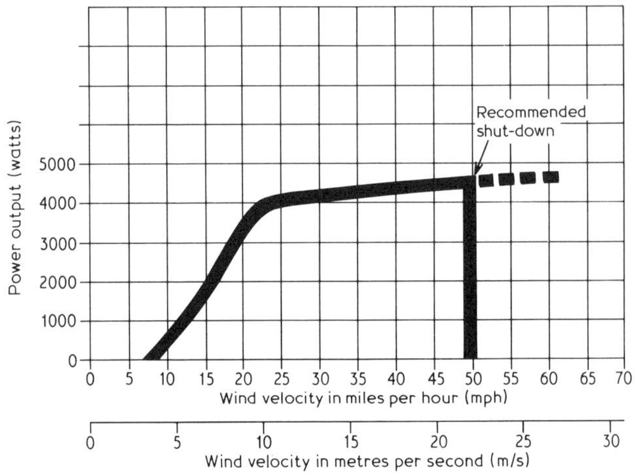

PERFORMANCE AND OPERATIONAL DATA

Rated electrical power	4000 watt
Rated output voltage	32/120/200 volt d.c.
Cut-in wind speed	3–3.5 m/s
Rated wind speed	10.3 m/s
Shut-down wind speed (recommended)	22.3 m/s
Survival wind speed	44.6 m/s
Shaft speed at rated output	285 rev/min
Maximum shaft speed	300 rev/min

PHYSICAL DATA

Turbine

Type: Horizontal axis, upwind with tail vane

Dimensions

diameter 4.58 m

Blades

Construction: Aircraft quality sitka spruce with stainless-steel leading edge

Number: 3

Control: Centrifugally operated blade pitch control

Generator: d.c. generator

Mechanical transmission: Direct coupling to turbine

System weight

net 272 kg

PRODUCTION

First production: 1979 (activity began earlier through partial rebuilding of Jacobs machines)

Production quantity to date: About 5 (October 1979)

MISCELLANEOUS

Two year warranty on materials and workmanship.

Semi-annual maintenance, comprising lubrication of five zurk fittings; blades to be refinished every five years.

Shut-down achieved by turning turbine into plane of tail vane and wind (manual operation).

Unarco-Rohn towers are offered.

VARIANT

Sky Hawk IV 'Sea Breeze', modified especially for marine environment.

Independent Energy Systems Inc.,
6043 Sterrettania Rd., Fairview,
Pennsylvania, 16415
USA
Telephone: (814) 833-3567

Iwanaka Inelco–M1

PERFORMANCE AND OPERATIONAL DATA

Rated electrical power	200 watt
Rated output voltage	24 volt d.c.
Cut-in wind speed	4 m/s
Rated wind speed	12 m/s
Shut-down wind speed	20 m/s
Survival wind speed	30 m/s
Maximum shaft speed	150 rev/min

PHYSICAL DATA

Turbine

Type: Vertical axis, straight blade type

Dimensions

height 2 m

diameter 2 m

Blades

Number: 4

Control: Fixed pitch turbine

Generator: Rectified a.c. generator

Mechanical transmission: Direct coupling to turbine

PRODUCTION

First production: May 1976

Production quantity to date: 3 (June 1981)

MISCELLANEOUS

Annual maintenance requirement.

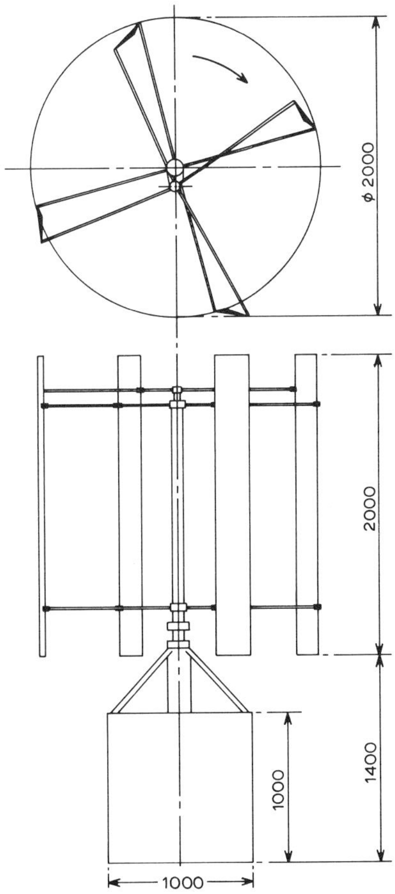

All dimensions in mm

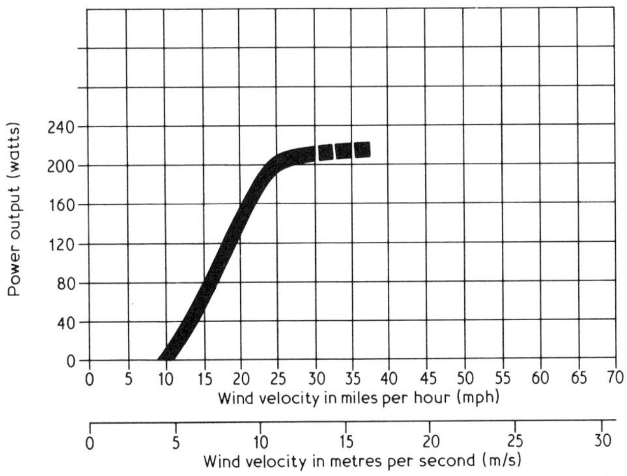

Iwanaka Electric Works Ltd,
3-5-8 Oyama, Sagamihara-Shi,
Kanagawa-Ken 229,
Japan

Jay Carter Model 25

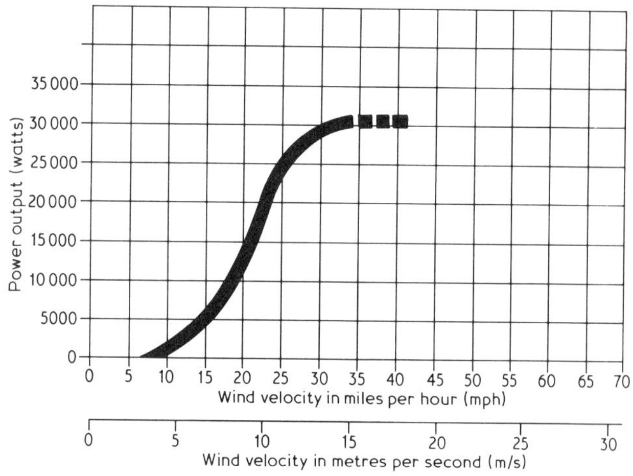

PERFORMANCE AND OPERATIONAL DATA

Rated electrical power	25 000 watt
Rated output voltage	220/440 volt a.c.
(operated in parallel with 3-phase 60 Hz network)	
Cut-in wind speed	3 m/s
Rated wind speed	11.6 m/s
Shut-down wind speed	none
Survival wind speed (actually experienced)	45 m/s
Shaft speed at rated output	120 rev/min

PHYSICAL DATA

Turbine

Type: Horizontal axis, downwind turbine operating freely in yaw

Dimensions

diameter 9.75 m

Blades

Construction: Fibreglass (continuous filament wound) and PVC foam, coated with polyester pigmented resin

Number: 2

Control: Fixed pitch turbine

Generator: 3-phase induction generator

Mechanical transmission: 15:1 ratio two-stage helical gearbox

System weight: 910 kg

PRODUCTION

First production: 1980

Production quantity to date: 100 (September 1982)

MISCELLANEOUS

Expected maintenance includes inspection every three months, gearbox oil change every 75 000 kW h, blades re-painted and leading edge renewed every 3–5 years.

18 m high galvanized pole supported with four guy wires.

Jay Carter Enterprises Inc.,
PO Box 684, Burkburnett,
Texas 76354,
USA
Telephone: (817) 569 2238

Jydsk Vindkraft
JV 55 A15

PERFORMANCE AND OPERATIONAL DATA

Rated electrical power 15 000 watt
Rated output voltage 380 volt
(operated in parallel with 3-phase 50 Hz network)
Cut-in wind speed 4.5 m/s
Rated wind speed 9.5 m/s
Survival wind speed 25 m/s
Maximum shaft speed 116 rev/min

PHYSICAL DATA

Turbine
Type: Horizontal axis
Dimensions
diameter 8.4 m
Blades
Construction: Reinforced glass fibre
Number: 3
Control: Variable pitch turbine (using servo motor)
Generator: Induction generator
Mechanical transmission: Speed increasing gearbox
System weight: 1200 kg (tower head)

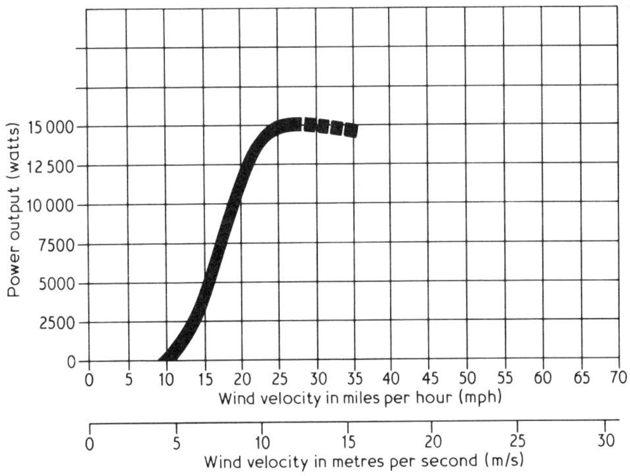

PRODUCTION

First production: October 1977
Production quantity to date: 5 (January 1980)

MISCELLANEOUS

Semi-annual maintenance comprising ordinary greasing and lubrication.
12 m, 15 m or 18 m guyed tubular tower.

VARIANTS

Other voltages and frequencies available.

Jydsk Vindkraft,
DK 8763 Rask Mølle,
Denmark
Telephone: (05) 67 8928

Kedco 1200

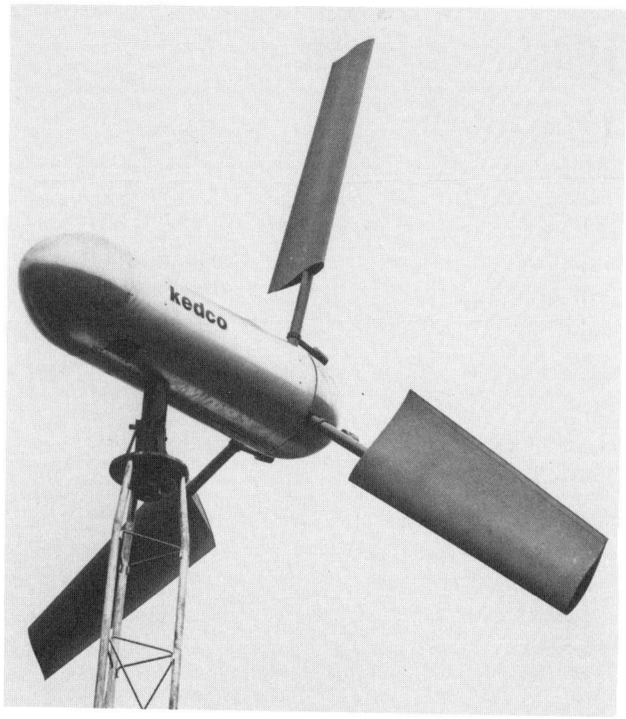

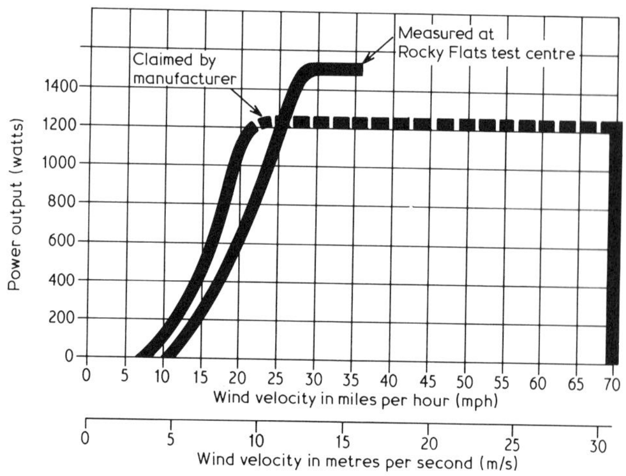

PERFORMANCE AND OPERATIONAL DATA

Rated electrical power	1200 watt
Rated output voltage	12 volt d.c.
Cut-in wind speed	3.1 m/s
Rated wind speed	9.3–9.8 m/s
Shut-down wind speed	31.1 m/s
Survival wind speed	49.1 m/s
Shaft speed at rated output	300 rev/min
Maximum shaft speed	300 rev/min

PHYSICAL DATA

Turbine
Type: Horizontal axis, downwind, operating freely in yaw
Dimensions
diameter 3.6 m
Blades
Construction: Aluminium 2024-T3
Number: 3
Control: Centrifugally activated pitch control
Generator: a.c. generator
Mechanical transmission: 8.76:1 ratio gearbox
System weight
net 92 kg
gross 160 kg

PRODUCTION

First production: 1976/77
Production quantity to date: 25 of the 1200 and 1600 series (September 1979)

MISCELLANEOUS

All parts plated, painted or anodized.
Automatic vibration sensing cut-off.
One year warranty covering parts and labour.
Annual maintenance comprising system inspection, checking fluid levels.
Handbook available.

Kedco Inc.,
9016 Aviation Boulevard,
Inglewood, California 90301,
USA
Telephone: (213) 776 6636

Kedco 1205

PERFORMANCE AND OPERATIONAL DATA

Rated electrical power	1900 watt
Rated output voltage	24 volt d.c.
Cut-in wind speed	3–4 m/s
Rated wind speed	10.7–11.2 m/s
Shut-down wind speed	31.1 m/s
Survival wind speed	49.1 m/s
Shaft speed at rated output	300 rev/min
Maximum shaft speed	300 rev/min

PHYSICAL DATA

Turbine

Type: Horizontal axis, downwind, operating freely in yaw

Dimensions

diameter 3.6 m

Blades

Construction: Aluminium 2024-T3

Number: 3

Control: Centrifugally activated pitch control

Generator: a.c. generator

Mechanical transmission: 8.76:1 ratio gearbox

System weight

net 92 kg

gross 160 kg

PRODUCTION

First production: 1976/77

Production quantity to date: 25 of the 1200 and 1600 series (September 1979)

MISCELLANEOUS

All parts plated, painted or anodized.
Automatic vibration sensing cut-off.
One year warranty covering parts and labour.
Annual maintenance comprising system inspection, checking fluid levels.
Handbook available.

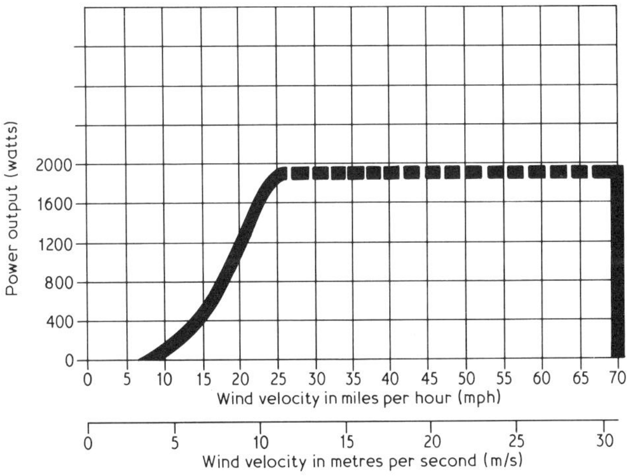

Kedco Inc.,
9016 Aviation Boulevard,
Inglewood, California 90301,
USA
Telephone: (213) 776 6636

Kedco 1210

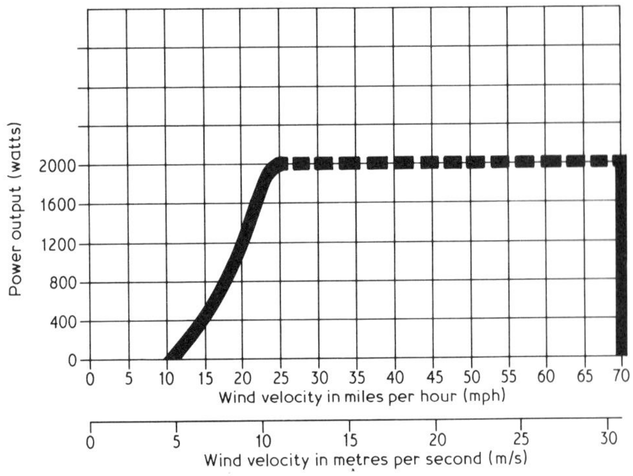

PERFORMANCE AND OPERATIONAL DATA

Rated electrical power	2000 watt
Rated output voltage	180 volt d.c.
Cut-in wind speed	4.4–5.3 m/s
Rated wind speed	11.1–11.6 m/s
Shut-down wind speed	31.1 m/s
Survival wind speed	49.1 m/s
Shaft speed at rated output	300 rev/min
Maximum shaft speed	300 rev/min

PHYSICAL DATA

Turbine
Type: Horizontal axis, downwind, operating freely in yaw
Dimensions
diameter 3.6 m
Blades
Construction: Aluminium 2024-T3
Number: 3
Control: Centrifugally activated pitch control
Generator: d.c. generator
Mechanical transmission: 8.76:1 ratio gearbox
System weight
net 114 kg
gross 182 kg

PRODUCTION

First production: 1976/77
Production quantity to date: 25 of the 1200 and 1600 series (September 1979)

MISCELLANEOUS

All parts plated, painted or anodized.
Automatic vibration sensing cut-off.
One year warranty covering parts and labour.
Annual maintenance comprising system inspection, checking fluid levels.
Handbook available.
This machine is designed for network operation through a synchronous inverter.

Kedco Inc.,
9016 Aviation Boulevard,
Inglewood, California 90301,
USA
Telephone: (213) 776 6636

Kedco 1600

PERFORMANCE AND OPERATIONAL DATA

Rated electrical power	1200 watt
Rated output voltage	12 volt d.c.
Cut-in wind speed	2.6–3.5 m/s
Rated wind speed	7.1–7.6 m/s
Shut-down wind speed	26.7 m/s
Survival wind speed	49.1 m/s
Shaft speed at rated output	250 rev/min
Maximum shaft speed	250 rev/min

PHYSICAL DATA

Turbine
Type: Horizontal axis, downwind, operating freely in yaw
Dimensions
diameter 4.88 m
Blades
Construction: Aluminium 2024-T3
Number: 3
Control: Centrifugally activated pitch control
Generator: a.c. generator
Mechanical transmission: 8.76:1 ratio gearbox
System weight
net 98 kg
gross 166 kg

PRODUCTION

First production: 1976/77
Production quantity to date: 25 of the 1200 and 1600 series (September 1979)

MISCELLANEOUS

All parts plated, painted or anodized.
Automatic vibration sensing cut-off.
One year warranty covering parts and labour.
Annual maintenance comprising system inspection, checking fluid levels.
Handbook available.

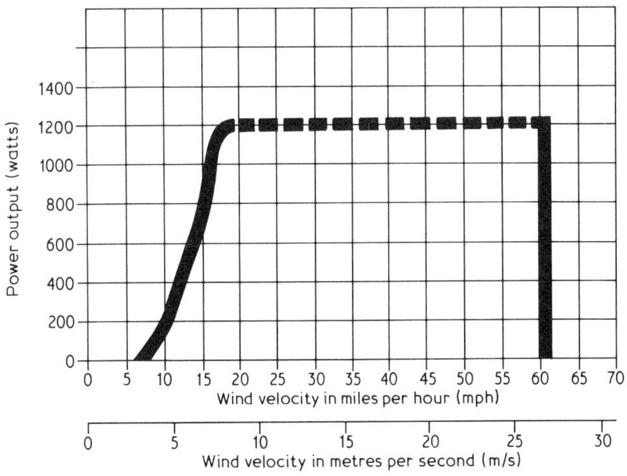

Kedco Inc.,
9016 Aviation Boulevard,
Inglewood, California 90301,
USA
Telephone: (213) 776 6636

Kedco 1605

PERFORMANCE AND OPERATIONAL DATA

Rated electrical power	1900 watt
Rated output voltage	24 volt d.c.
Cut-in wind speed	4–4.9 m/s
Rated wind speed	8.9–9.3 m/s
Shut-down wind speed	26.7 m/s
Survival wind speed	49.1 m/s
Shaft speed at rated output	250 rev/min
Maximum shaft speed	250 rev/min

PHYSICAL DATA

Turbine
Type: Horizontal axis, downwind, operating freely in yaw
Dimensions
diameter 4.88 m
Blades
Construction: Aluminium 2024-T3
Number: 3
Control: Centrifugally activated pitch control
Generator: a.c. generator
Mechanical transmission: 8.76:1 ratio gearbox
System weight
net 98 kg
gross 166 kg

PRODUCTION

First production: 1976/77
Production quantity to date: 25 of the 1200 and 1600 series (September 1979)

MISCELLANEOUS

All parts plated, painted or anodized.
Automatic vibration sensing cut-off.
One year warranty covering parts and labour.
Annual maintenance comprising system inspection, checking fluid levels.
Handbook available.

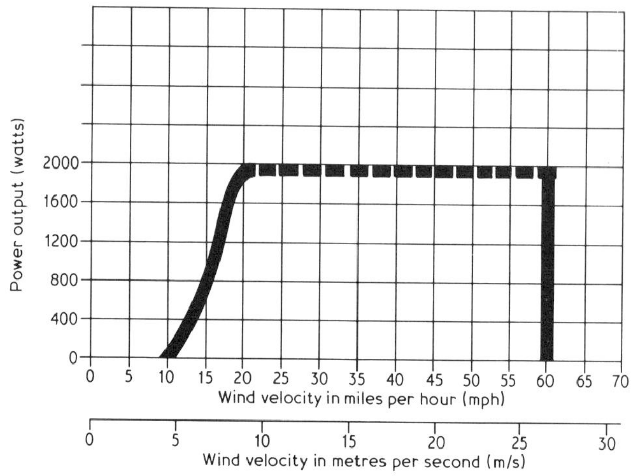

Kedco Inc.,
9016 Aviation Boulevard,
Inglewood, California 90301,
USA
Telephone: (213) 776 6636

Kedco 1610

PERFORMANCE AND OPERATIONAL DATA

Rated electrical power	2000 watt
Rated output voltage	180 volt d.c.
Cut-in wind speed	4–4.9 m/s
Rated wind speed	9.3–9.8 m/s
Shut-down wind speed	26.7 m/s
Survival wind speed	49.1 m/s
Shaft speed at rated output	250 rev/min
Maximum shaft speed	250 rev/min

PHYSICAL DATA

Turbine
Type: Horizontal axis, downwind, operating freely in yaw
Dimensions
diameter 4.88 m
Blades
Construction: Aluminium 2024-T3
Number: 3
Control: Centrifugally activated pitch control
Generator: d.c. generator
Mechanical transmission: 8.76:1 ratio gearbox
System weight
net 121 kg
gross 189 kg

PRODUCTION

First production: 1976/77
Production quantity to date: 25 of the 1200 and 1600 series (September 1979)

MISCELLANEOUS

All parts plated, painted or anodized.
Automatic vibration sensing cut-off.
One year warranty covering parts and labour.
Annual maintenance comprising system inspection, checking fluid levels.
Handbook available.
This machine is designed for network operation through a synchronous inverter.

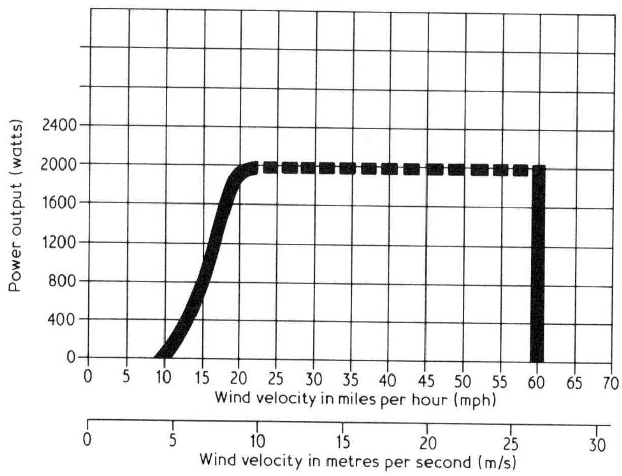

Kedco Inc.,
9016 Aviation Boulevard,
Inglewood, California 90301,
USA
Telephone: (213) 776 6636

Kedco
1620

PERFORMANCE AND OPERATIONAL DATA

Rated electrical power	3000 watt
Rated output voltage	180 volt d.c.
Cut-in wind speed	4.4–5.3 m/s
Rated wind speed	11.1–11.6 m/s
Shut-down wind speed	26.7 m/s
Survival wind speed	49.1 m/s
Shaft speed at rated output	250 rev/min
Maximum shaft speed	250 rev/min

PHYSICAL DATA

Turbine
Type: Horizontal axis, downwind, operating freely in yaw
Dimensions
diameter 4.88 m
Blades
Construction: Aluminium 2024-T3
Number: 3
Control: Centrifugally activated pitch control
Generator: d.c. generator
Mechanical transmission: 8.76:1 ratio gearbox
System weight
net 133 kg
gross 201 kg

PRODUCTION

First production: 1976/77
Production quantity to date: 25 of the 1200 and 1600 series (September 1979)

MISCELLANEOUS

All parts plated, painted or anodized.
Automatic vibration sensing cut-off.
One year warranty covering parts and labour.
Annual maintenance comprising system inspection, checking fluid levels.
Handbook available.
This machine is designed for network operation through a synchronous inverter.

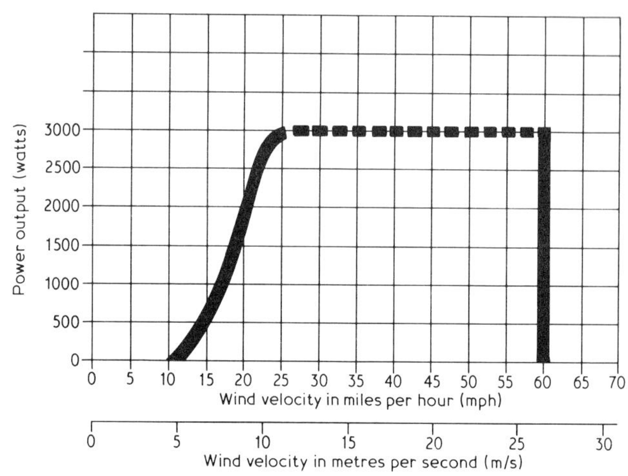

Kedco Inc.,
9016 Aviation Boulevard,
Inglewood, California 90301,
USA
Telephone: (213) 776 6636

Lubing
M 022-3 G 024-400

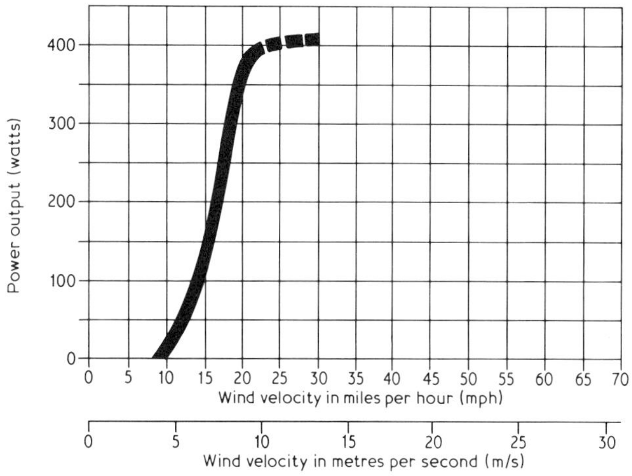

PERFORMANCE AND OPERATIONAL DATA

Rated electrical power	400 watt
Rated output voltage	12/24 volt d.c.
Cut-in wind speed	3.5–4 m/s
Rated wind speed	10 m/s
Shut-down wind speed	none
Maximum shaft speed	600 rev/min

PHYSICAL DATA

Turbine

Type: Horizontal axis, downwind, operating freely in yaw

Dimensions

diameter 2.2 m

Blades

Construction: Expanded polystyrene, pressed and covered with epoxy fibreglass

Number: 6

Control: Small (starting) blades fixed pitch; large blades variable pitch (centrifugally activated)

Generator: Brushless a.c. generator

Mechanical transmission: Two-stage gearbox

PRODUCTION

Production discontinued 1981/82

MISCELLANEOUS

Guyed tubular tower offered in 3 m sections with 6 m minimum. Coding above refers to tower head unit only; M 022-3-6 G 024-400 would for instance include a 6 m tower.

9 m and 12 m towers are offered.

Lubing Maschinenfabrik,
Postfach 110,
D-2847 Barnstorf,
West Germany
Telephone: (05442) 625/6/7
Telex: 09-41212

Matsushita FY-04 WG1

PERFORMANCE AND OPERATIONAL DATA

Rated electrical power	1000 watt
Rated output voltage	24 volt d.c.
Cut-in wind speed	2.5 m/s
Rated wind speed	8.5 m/s
Shut-down wind speed	30 m/s
Survival wind speed	60 m/s
Maximum shaft speed	250 rev/min

PHYSICAL DATA

Turbine
Type: Horizontal axis, upwind with tail vane
Dimensions
diameter 4.0 m
Blades
Construction: FRP resin
Number: 2
Control: Centrifugally operated blade pitch control
Generator: a.c. generator
Mechanical transmission: Speed increasing gearbox
System weight: 350 kg (power generator)

PRODUCTION

First production: 1978
Production quantity to date: 4 (September 1979)

MISCELLANEOUS

One year warranty period covering damage or malfunction arising through defect of design or manufacture, but excluding damage arising from operation outside stated limits.
Annual maintenance, comprising gearbox oil change, checking for tightness of nuts and bolts and electrical connections, checking wear and freedom of brushes, inspection of brake and checking for corrosion or unusual operational noise.
10 m steel pole offered as tower.
Ambient temperature limits −40°C to +60°C (−10°C to +50°C while operating).

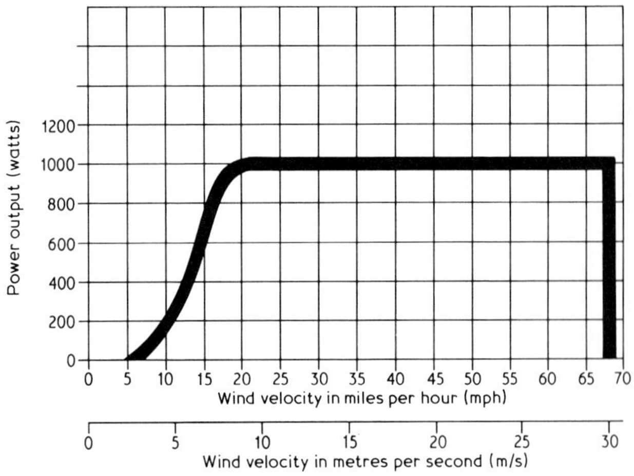

Matsushita Seiko Co. Ltd,
Takagi-cho, Kasugai-shi,
Aichi 486, Japan

Matsushita FY-08 WG4

PERFORMANCE AND OPERATIONAL DATA

Rated electrical power	2000 watt
Rated output voltage	24 volt d.c.
Cut-in wind speed	2.5 m/s
Rated wind speed	6.5 m/s
Shut-down wind speed	20 m/s
Survival wind speed	60 m/s
Maximum shaft speed	70 rev/min

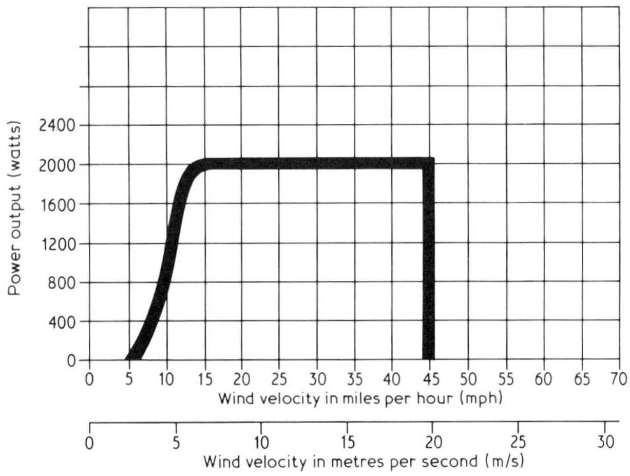

PHYSICAL DATA

Turbine

Type: Horizontal axis, upwind with tail vane

Dimensions

diameter 8.8 m

Blades

Construction: FRP resin

Number: 2

Control: Blade pitch controlled by motor cylinder

Generator: a.c. generator

Mechanical transmission: Speed increasing gearbox

System weight: 2000 kg (power generator)

PRODUCTION

First production: 1977

Production quantity to date: 2 (September 1979)

MISCELLANEOUS

One year warranty period covering damage or malfunction arising through defect of design or manufacture, but excluding damage arising from operation outside stated limits.

Annual maintenance, comprising gearbox oil change, checking for tightness of nuts and bolts and electrical connections, checking wear and freedom of brushes, inspection of brake and checking for corrosion or unusual operational noise.

10 m truss tower offered.

Ambient temperature limits −40°C to +60°C (−10°C to +50°C while operating).

Matsushita Seiko Co. Ltd.,
Takagi-cho, Kasugai-shi,
Aichi 486,
Japan

Megatech WIP-A1

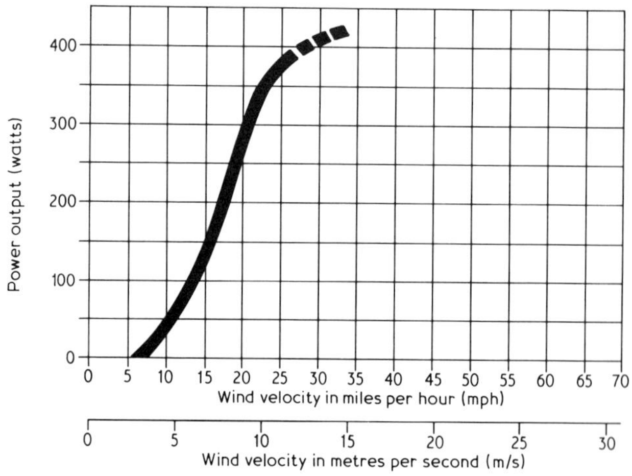

PERFORMANCE AND OPERATIONAL DATA

Rated electrical power	400 watt
Rated output voltage	12 volt d.c.
Cut-in wind speed	2.7 m/s
Rated wind speed	12.0 m/s
Survival wind speed	35–45 m/s (with manual brake applied)
Maximum shaft speed	1000 rev/min

PHYSICAL DATA

Turbine

Type: Horizontal axis, upwind with tail vane

Dimensions

diameter 1.83 m

Blades

Construction: Wood coated with fibreglass resin

Number: 2

Control: Fixed pitch turbine; high thrust turns turbine sideways to the wind

Generator: Internally excited d.c. generator

Mechanical transmission: Direct coupling to turbine

System weight: 113 kg (approx)

PRODUCTION

First production: 1976

Production quantity to date: In the order of 100s

MISCELLANEOUS

Intended for educational purposes; offered with a software package.

One year warranty on material and workmanship.

Annual maintenance comprising lubrication and brush check.

Megatech International Inc.,
29 Cook Street, Billerica,
MA 01866,
USA
Telephone: (617) 273 1900

Mehrkam Energy Development Company Model 440

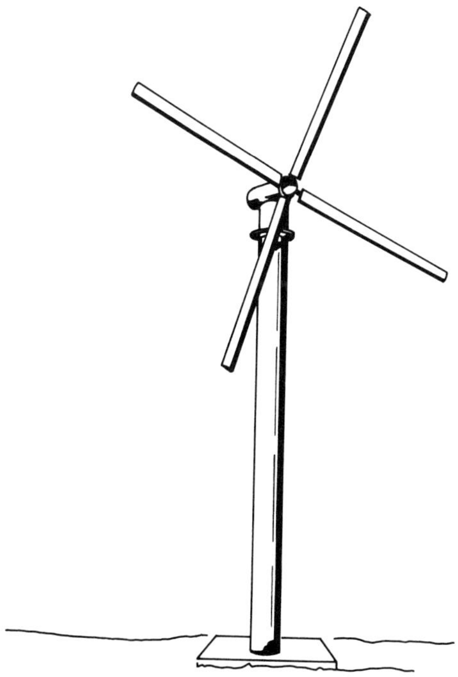

PERFORMANCE AND OPERATIONAL DATA

Rated electrical power	40000 watt
Rated output voltage	240 volt a.c.
(operated in parallel with 60 Hz power network)	
Cut-in wind speed	2.2 m/s
Rated wind speed	11.2 m/s
Shut-down wind speed	17.8 m/s
Survival wind speed	53.5 m/s
Shaft speed at rated output	40 rev/min (approx.)
Maximum shaft speed	60 rev/min

PHYSICAL DATA

Turbine
Type: Horizontal axis, downwind type
Dimensions
diameter 11.0 or 11.6 m

Blades
Construction: Aluminium, T-6 (extruded)
Number: 6

Control: Automatic yaw control
Generator: Induction generator
Mechanical transmission: 20:1 ratio gearbox
System weight
net 3175 kg

PRODUCTION

First production: 1976
Production quantity to date: Combined production of all Mehrkam machines is about 50 (June 1981)

MISCELLANEOUS

Two year warranty on materials and workmanship.
Maintenance comprises greasing every 6 months (an automatic lubricator can be included), oil change every two years.
Two electrically operated friction brakes are included.

VARIANT

Synchronous generator with inverter is offered.

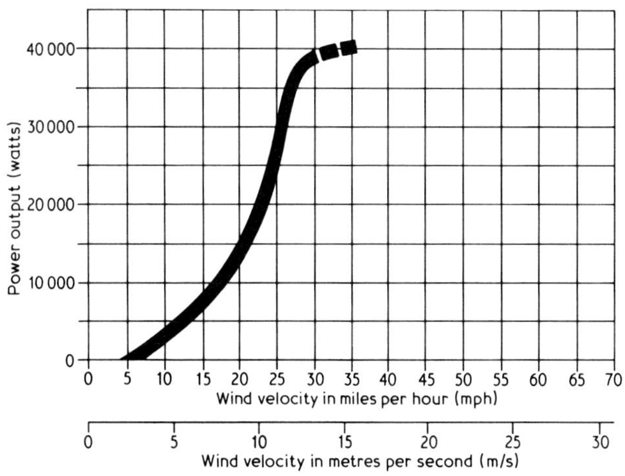

Mehrkam Energy Development Co.,
RD2, Box 179E, Hamburg,
Pennsylvania 19526,
USA
Telephone: (215) 562 8856

Millville 15-2

PERFORMANCE AND OPERATIONAL DATA

Rated electrical power 15 000 watt
Rated output voltage 240 volt a.c.
(operated in parallel with single or 3 phase, 60 Hz power network)
Cut-in wind speed 4 m/s
Rated wind speed 10.3 m/s
Shut-down wind speed 27 m/s
Survival wind speed 36 m/s
Shaft speed at rated output 80 rev/min
Maximum shaft speed 85 rev/min

PHYSICAL DATA

Turbine
Type: Horizontal axis, upwind with tail vane
Dimensions
diameter 8.8 m
Blades
Construction: Aluminium
Number: 3
Control: Mechanical blade pitch change; in high winds turbine turns into plane of turbine and tail vane
Generator: Induction generator
Mechanical transmission: 24:1 ratio gearbox
System weight
net 386 kg

PRODUCTION

First production: 1977
Production quantity to date: About 35 (October 1979)

MISCELLANEOUS

Offered with 12.2 m, 18.3 m, 24.4 m or 30.5 m Rohn type SSV towers.
One year warranty.
Semi-annual maintenance comprising oil check and system inspection.
Mechanical disc brake.
Complete operation moved from Millville, California to Honolulu in November 1981.

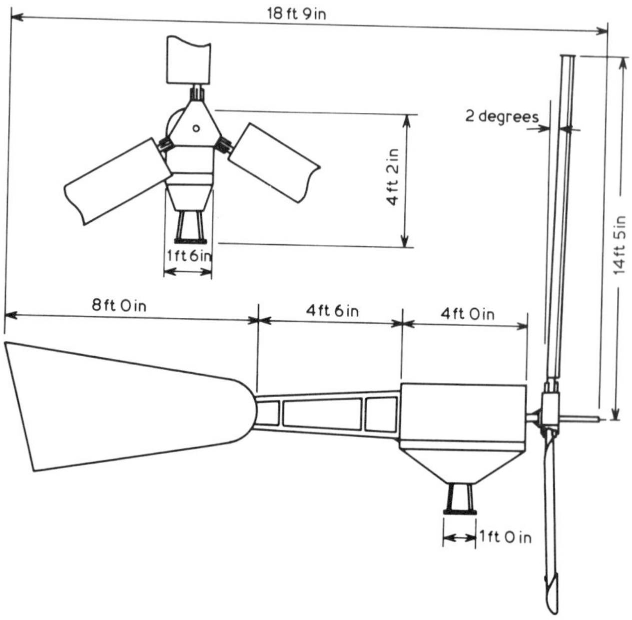

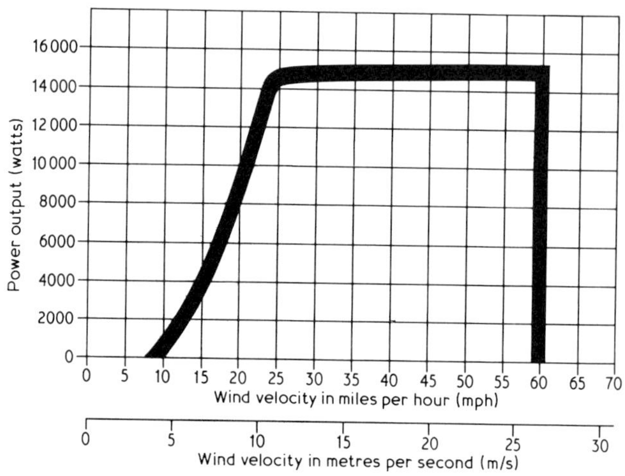

Millville Hawaii Windmills Inc.,
3028 Valena Street,
Honolulu,
Hawaii 96819
USA
Telephone: (808) 836 1180

Nordtank
10 kW

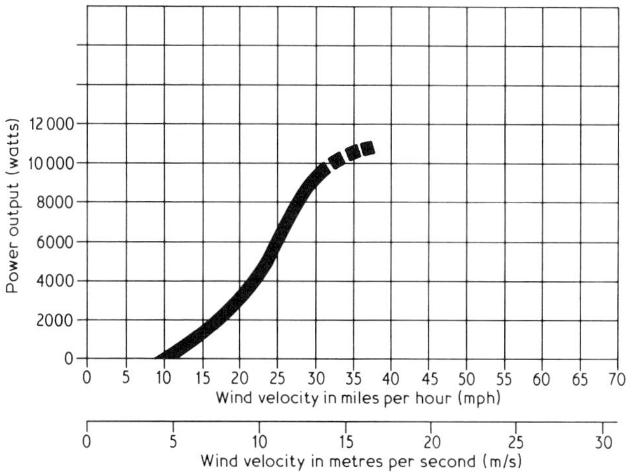

PERFORMANCE AND OPERATIONAL DATA

Rated electrical power	10 000 watt
Rated output voltage	380 volt a.c.
(operated in parallel with 3 phase, 50 Hz network)	
Cut-in wind speed	4 m/s
Rated wind speed	14 m/s
Shut-down wind speed	none
Survival wind speed	50 m/s
Shaft speed at rated output	80 rev/min

PHYSICAL DATA

Turbine
Type: Horizontal axis, downwind type
Dimensions
diameter 8.5 m
Blades
Construction: Glassfibre reinforced polyester
Number: 3
Control: Fixed pitch turbine
Generator: Induction generator
System weight: 2000 kg

PRODUCTION

First production: 1980
Production quantity to date: 10 (June 1981)

MISCELLANEOUS

Two year guarantee with free service.
17 m tubular tower.

Nordtank,
T. Rørbaek Jensen A/S,
Nyballevej 8, DK-8444 Balle,
Denmark
Telephone: (06) 33 7200

Nordtank 22 kW

PERFORMANCE AND OPERATIONAL DATA

Rated electrical power	22 000 watt
Rated output voltage	380 volt a.c.
(operated in parallel with 3 phase, 50 Hz network)	
Cut-in wind speed	4 m/s
Rated wind speed	14 m/s
Shut-down wind speed	none
Survival wind speed	50 m/s

PHYSICAL DATA

Turbine
Type: Horizontal axis, upwind type
Dimensions
diameter 11.0 m
Blades
Construction: Glassfibre reinforced polyester
Number: 3
Control: Fixed pitch turbine
Generator: Two individual induction generators, 22 kW for high wind speed, 7.5 kW for low wind speed
System weight: 4500 kg

PRODUCTION

First production: 1980
Production quantity to date: 10 (June 1981)

MISCELLANEOUS

Two year guarantee with free service.
18 m tubular tower.

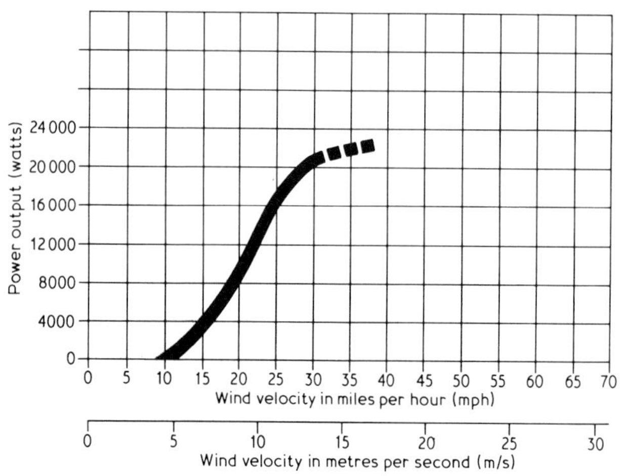

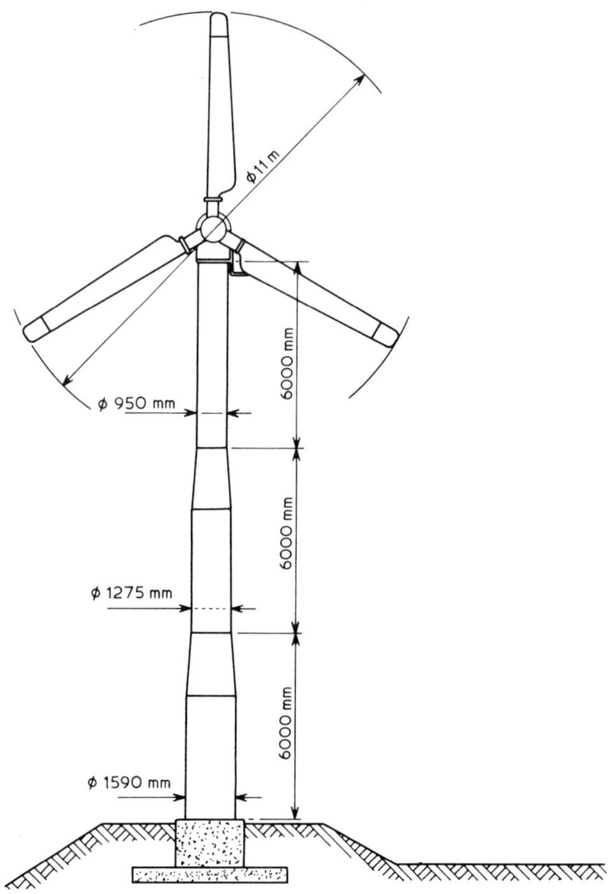

Nordtank,
T. Rørbaek Jensen A/S,
Nyballevej 8, DK-8444 Balle,
Denmark
Telephone: (06) 33 7200

Nordtank
30 kW

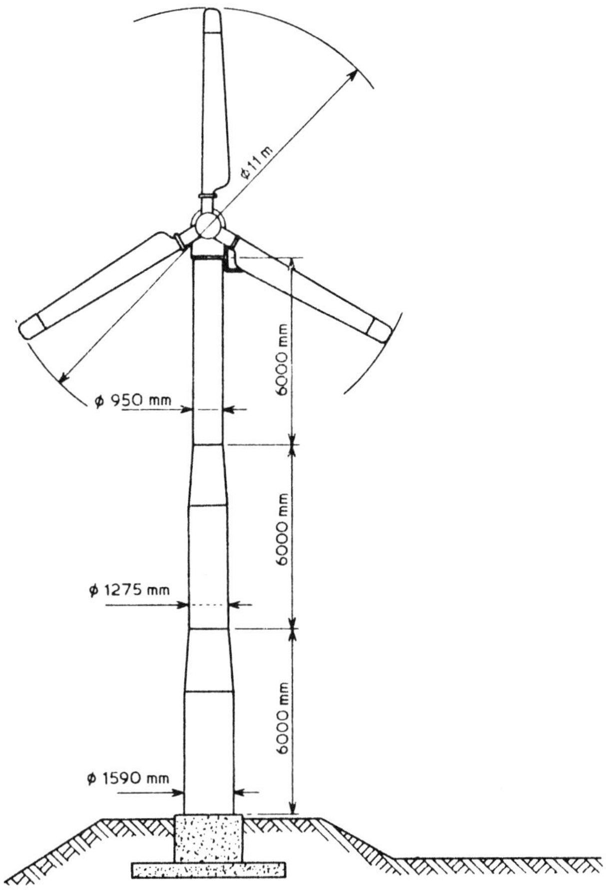

PERFORMANCE AND OPERATIONAL DATA

Rated electrical power	30 000 watt
Rated output voltage	380 volt a.c.
(operated in parallel with 3 phase, 50 Hz network)	
Cut-in wind speed	4 m/s
Rated wind speed	14 m/s
Shut-down wind speed	none
Survival wind speed	50 m/s

PHYSICAL DATA

Turbine

Type: Horizontal axis, upwind type

Dimensions

diameter 11.0 m

Blades

Construction: Glassfibre reinforced polyester

Number: 3

Control: Fixed pitch turbine

Generator: Two individual induction generators, 30 kW for high wind speed, 7.5 kW for low wind speed

System weight: 4500 kg

PRODUCTION

First production: 1980

Production quantity to date: 15 (June 1981)

MISCELLANEOUS

Two year guarantee with free service.

18 m tubular tower.

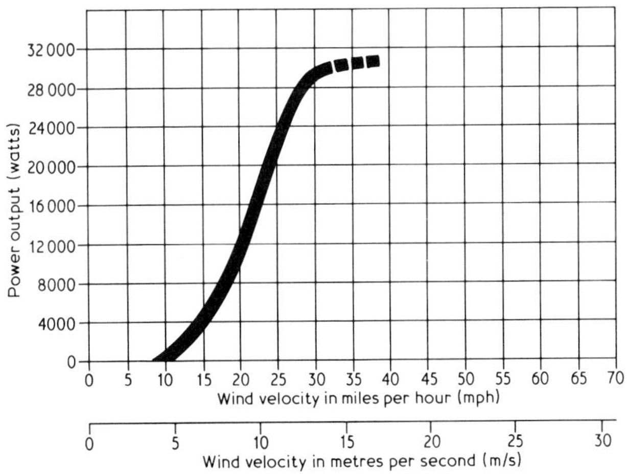

Nordtank,
T. Rørbaek Jensen A/S,
Nybellevej 8, DK-8444 Balle,
Denmark
Telephone: (06) 33 7200

Nordtank 45 kW

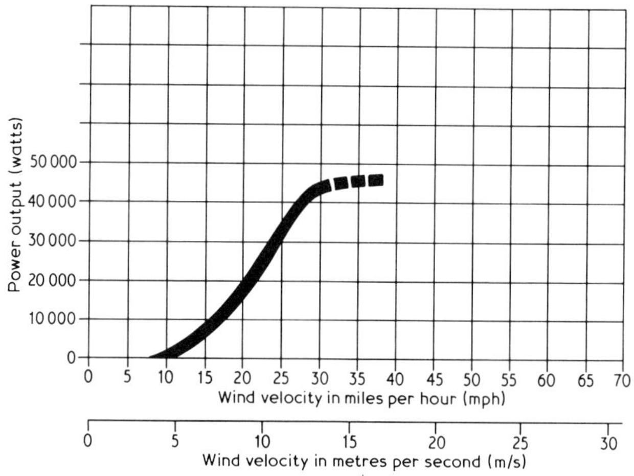

PERFORMANCE AND OPERATIONAL DATA

Rated electrical power	45 000 watt
Rated output voltage	380 volt a.c.
(operated in parallel with 3 phase, 50 Hz network)	
Cut-in wind speed	4 m/s
Rated wind speed	14 m/s
Shut-down wind speed	none
Survival wind speed	50 m/s

PHYSICAL DATA

Turbine
Type: Horizontal axis, upwind type
Dimensions
diameter 15.5 m
Blades
Construction: Glassfibre reinforced polyester
Number: 3
Control: Fixed pitch turbine
Generator: Two individual generators, 45 kW for high wind speed, 11 kW for low wind speed
System weight: 10 000 kg

PRODUCTION

First production: 1980
Production quantity to date: 10 (June 1981)

MISCELLANEOUS

Two year guarantee with free service.
18 m or 22 m tower.

Nordtank,
T. Rørbaek Jensen A/S,
Nyballevej 8, DK-8444 Balle,
Denmark
Telephone: (06) 33 7200

Nordtank
55 kW

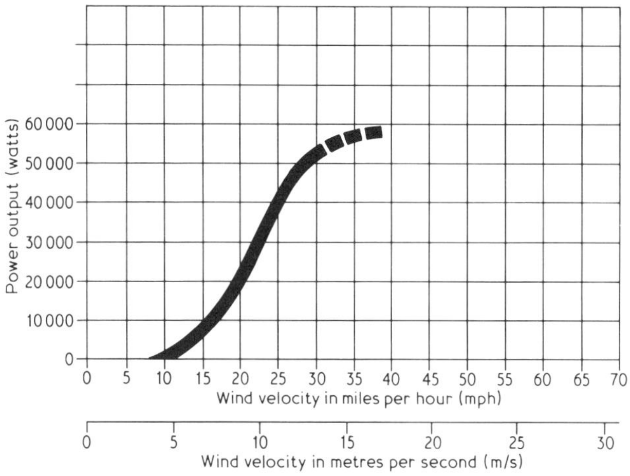

PERFORMANCE AND OPERATIONAL DATA

Rated electrical power 55 000 watt
Rated output voltage 380 volt a.c.
(operated in parallel with 3 phase, 50 Hz network)
Cut-in wind speed 4 m/s
Rated wind speed 14 m/s
Shut-down wind speed none
Survival wind speed 50 m/s

PHYSICAL DATA

Turbine
Type: Horizontal axis, upwind type
Dimensions
diameter 15.5 m
Blades
Construction: Glassfibre reinforced polyester
Number: 3
Control: Fixed pitch turbine
Generator: Two individual induction generators, 55 kW for high wind speed, 11 kW for low wind speed
System weight: 10 000 kg

PRODUCTION

First production: 1980
Production quantity to date: 25 (June 1981)

MISCELLANEOUS

Two year guarantee with free service.
18 m or 22 m tower.

Nordtank,
T. Rørbaek Jensen A/S,
Nybellevej 8, DK-8444 Balle,
Denmark
Telephone: (06) 33 7200

Nordvestsjaellands VM 10

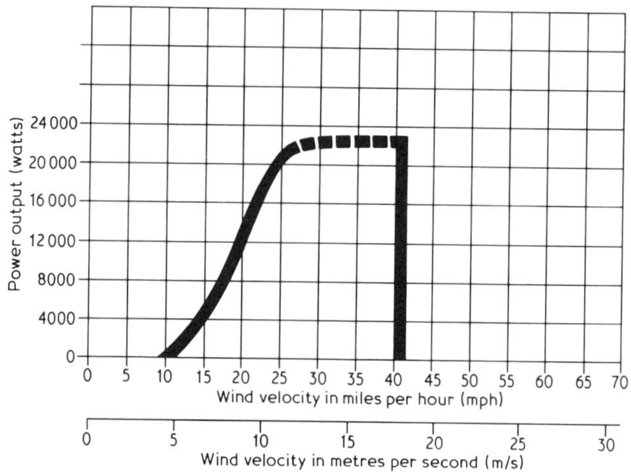

PERFORMANCE AND OPERATIONAL DATA

Rated electrical power	22 000 watt
Rated output voltage	380 volt a.c.
(operated in parallel with 3 phase, 50 Hz network)	
Cut-in wind speed	4–5 m/s
Rated wind speed	12 m/s
Shut-down wind speed	18 m/s
Survival wind speed	36 m/s
Shaft speed at rated output	67 rev/min

PHYSICAL DATA

Turbine
Type: Horizontal axis
Dimensions
diameter 10 m
Blades
Construction: Laminated wood
Number: 3
Control: Fixed pitch turbine
Generator: Two induction generators
Mechanical transmission: Speed increasing gearbox
System weight: 3500 kg

PRODUCTION

First production: 1978
Production quantity to date: 3 (August 1980)

MISCELLANEOUS

Annual maintenance requirement.

VARIANTS

Different generator combinations are offered.

Nordvestsjaellands
Energivaerksted,
Ringstedvej 189, 4300 Holbaek,
Denmark
Telephone: (03) 479135

North Wind Power HR2

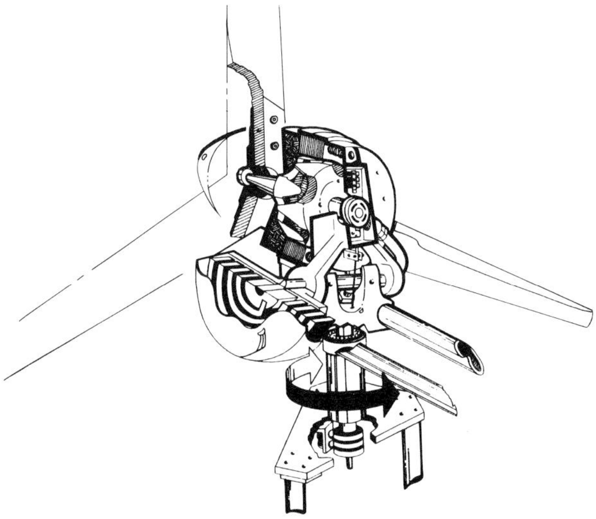

PERFORMANCE AND OPERATIONAL DATA

Rated electrical power	2300 watt
Rated output voltage	24/48/120 volt d.c.
Cut-in wind speed	3.6 m/s
Rated wind speed	9.0 m/s
Shut-down wind speed	47 m/s
Survival wind speed	73.7 m/s
Shaft speed at rated output	250 rev/min
Maximum shaft speed	400 rev/min

PHYSICAL DATA

Turbine
Type: Horizontal axis, upwind with tail vane
Dimensions
diameter 5 m
Blades
Construction: Wood composite
Number: 3
Control: Fixed pitch turbine
Generator: 3-phase Lundell-type a.c. generator (rectified)
Mechanical transmission: Direct coupling to turbine
System weight: 356 kg (tower head)

PRODUCTION

First production: 1979
Production quantity to date: 80 (June 1981)

MISCELLANEOUS

Annual maintenance requirement (one man-day).
One year warranty on parts and maintenance, limited three year guarantee.

VARIANTS

32/240 volt d.c. by special order. Cold weather option available to suit −40°C to −60°C range.

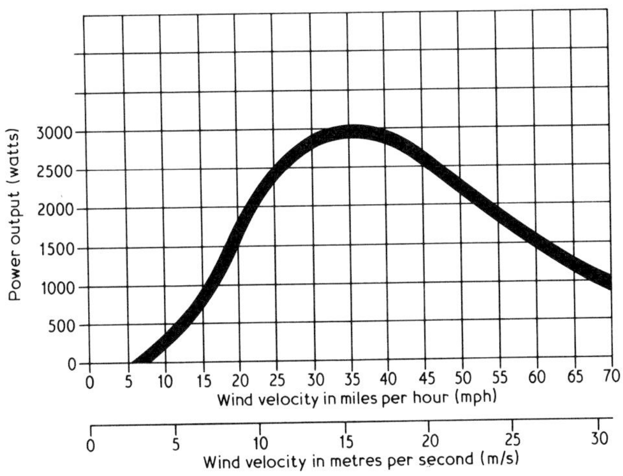

North Wind Power Co.,
P.O. Box 556, Moretown,
Vermont 05660,
USA
Telephone: (802) 496 2955
Telex: 957034

Panensa
5 kVA

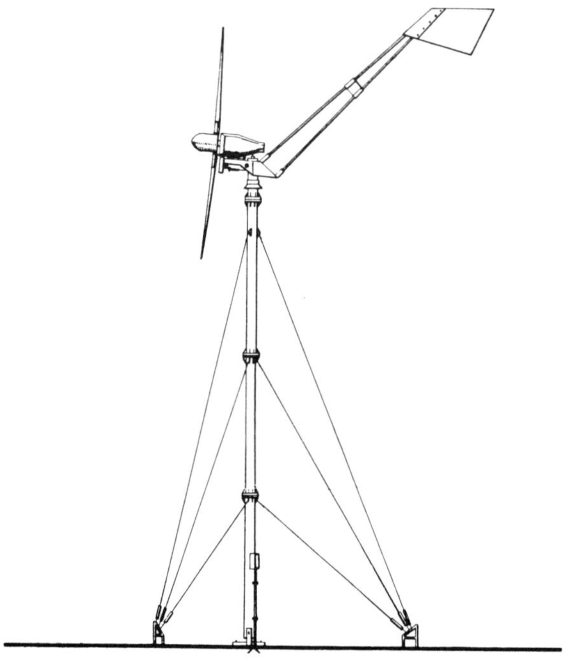

PERFORMANCE AND OPERATIONAL DATA

Rated electrical power	5000 watt
Rated output voltage	380–700 volt
Cut-in wind speed	4.0 m/s
Rated wind speed	10.0 m/s
Shut-down wind speed	10.0 m/s
Survival wind speed	41.7 m/s
Shaft speed at rated output	375 rev/min
Maximum shaft speed	450 rev/min

PHYSICAL DATA

Turbine
Type: Horizontal axis, upwind with tail vane
Dimensions
diameter 5 m
Blades
Construction: Resin reinforced fibreglass
Number: 2
Control: Variable pitch for starting and overspeed prevention
Generator: Permanent magnet a.c. generator
Mechanical transmission: Direct coupling to turbine
System weight: 800 kg

PRODUCTION

Production quantity to date: 10 (November 1979)

MISCELLANEOUS

One year warranty on labour and parts, 10 years spare parts.
Maintenance of brushes required (replacement every two years).

VARIANTS

The unit can be used to power an electrical water pumping system, or through a step-down (380/110 volt) transformer/rectifier for battery charging.

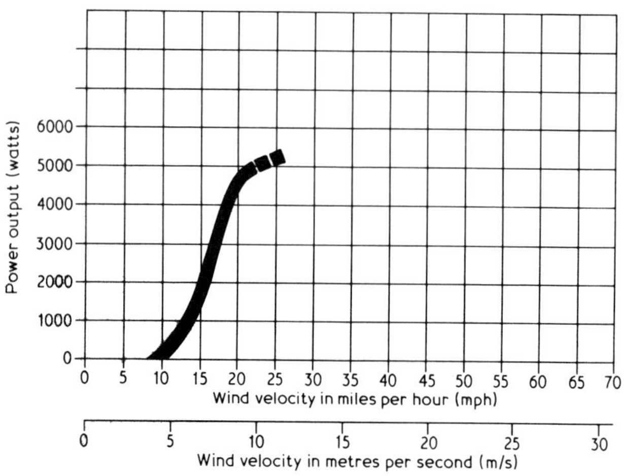

Panensa SA,
CH 2035 Corcelles/NE,
Switzerland
Telephone: (038) 31 34 34
Telex: 35140

Pinson Cycloturbine C2E3

PERFORMANCE AND OPERATIONAL DATA

Rated electrical power	5000 watt
Rated output voltage	various
Cut-in wind speed	3.1 m/s
Rated wind speed	13.4 m/s
Shut-down wind speed	22.4 m/s
Survival wind speed	54 m/s
Shaft speed at rated output	200 rev/min

PHYSICAL DATA

Turbine

Type: Vertical axis, straight blade type
Dimensions
diameter 4.89 m

Blades

Construction: 6061 T-6 aluminium extrusion leading edge, brake formed sheet trailing edge
Number: 3

Control: Blade pitch cyclically varied by cam and tail vane; internal centrifugally actuated shut-down mechanism
Mechanical transmission: Speed increasing gearbox
System weight: 200 kg (approx.)

PRODUCTION

First production: 1977
Production quantity to date: About 80 (June 1981)

MISCELLANEOUS

Annual maintenance, comprising greasing of bearings, tightening hardware, replacing drive belt.
One year warranty covering defects of materials or workmanship on rotor system, tower and generator.
Handbook available.
Earlier type used same blades but 3.65 m or 4.58 m diameter, rated 2000 watt.
a.c. generator offered for direct heating applications, d.c. generator for battery charging and network operation with inverter.

REPORTS

Rocky Flats Test Centre report:
Carr, M.J., Grotsky, V.K. and Sexton, J.H. *Pinson C2E Wind Turbine Generator Failure Analysis and Corrective Design Modification* RFP 3148/3533/80–16, March 1980.

Pinson Energy Corporation,
PO Box 7, Marston Mills,
Massachusetts 02648,
USA
Telephone: (617) 428 8535

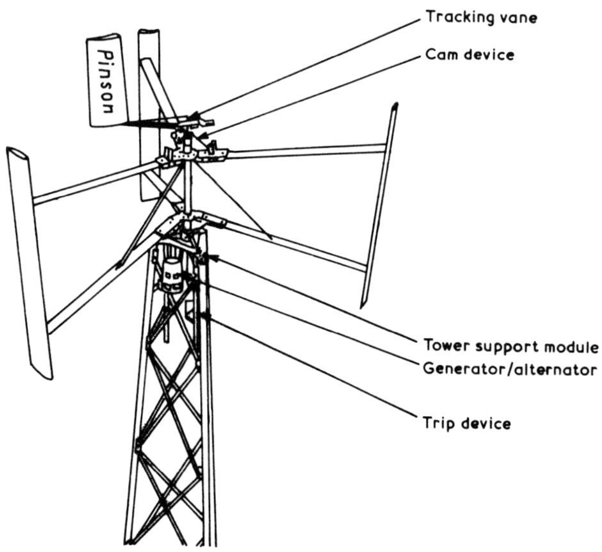

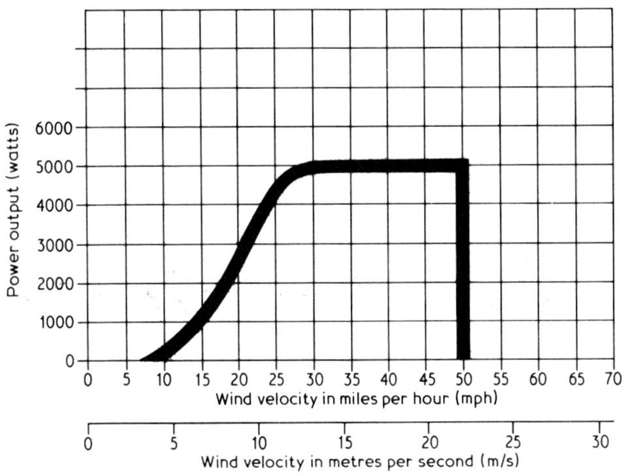

PI Specialist Engineers 6 metre

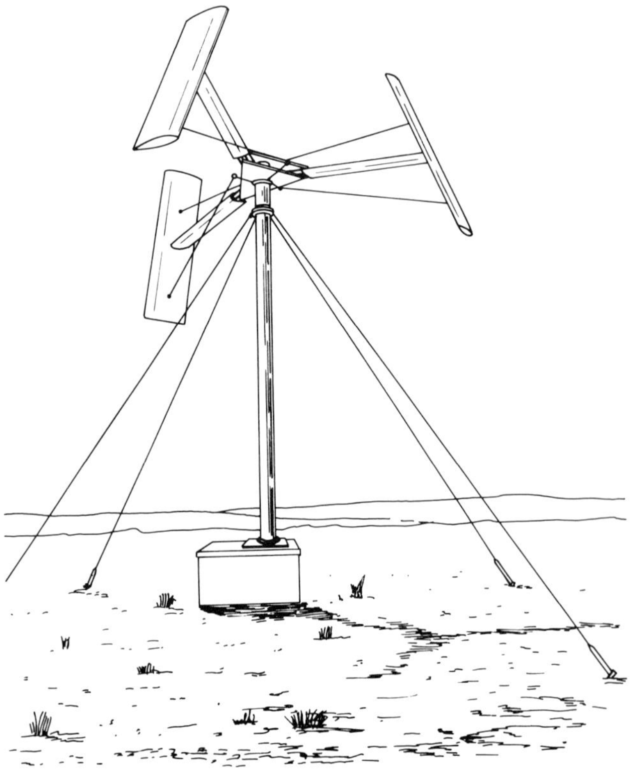

PERFORMANCE AND OPERATIONAL DATA

Rated electrical power	3200 watt
Rated output voltage	24 volt d.c.
Cut-in wind speed	4 m/s
Rated wind speed	15 m/s
Shut-down wind speed	none
Survival wind speed	over 50 m/s
Maximum shaft speed	120 rev/min

PHYSICAL DATA

Turbine

Type: Vertical axis, variable geometry, with hinged straight blades, profile NACA 0015. Turbine arms of steel box section with streamlined fairing

Dimensions
height 3 m
diameter 6 m

Blades

Construction: Steel skeleton with wood spars, clad with glass reinforced plastic

Number: 3

Control: Fixed blade pitch, but vertical attitude variable

Generator: Two d.c. generators

Mechanical transmission: Speed increasing gearbox

System weight
net 750 kg (including tower)

PRODUCTION

First production: 1979
Production quantity to date: 10 (June 1981)

MISCELLANEOUS

Standard tower height 7 m; guyed flanged steel tube.
One year warranty on materials and workmanship.
Annual maintenance required on gearbox and bearings.
Coning action of blades begins at 10 m/s wind speed.

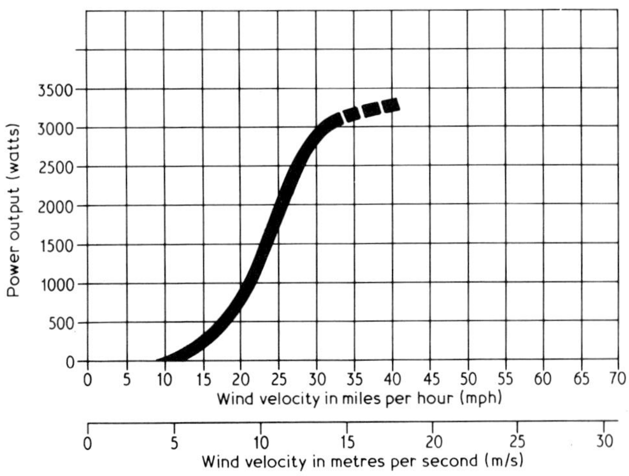

PI Specialist Engineers,
The Dean, Alresford,
Hants
Telephone: (0962 73) 3242
(0962 73) 4262
Telex: 477152 Picor G

Poulsen 30 kW

PERFORMANCE AND OPERATIONAL DATA

Rated electrical power	30 000 watt
Rated output voltage	380 volt a.c.
(operated in parallel with 3 phase, 50 Hz network)	
Cut-in wind speed	3 m/s
Rated wind speed	11 m/s
Shut-down wind speed	22 m/s
Survival wind speed	40 m/s
Shaft speed at rated output	120 rev/min
Maximum shaft speed	120 rev/min

PHYSICAL DATA

Turbine

Type: Horizontal axis, downwind, operating freely in yaw using turbine coning

Dimensions

diameter 13.0 m

Blades

Construction: Extruded aluminium

Number: 2

Control: Fixed pitch turbine

Generator: Two individual induction generators (5.5 kW for low wind speed, 30 kW for high wind speed)

Mechanical transmission: 12.5:1 ratio gearbox

System weight: 2000 kg

PRODUCTION

First production: 1978

Production quantity to date: 8 (June 1981)

MISCELLANEOUS

Low wind speed generator is 8 pole, turbine runs at 60 rev/min; high wind speed generator is 4 pole, turbine runs at 120 rev/min.

Generator and gearbox mounted near tower on inclined shaft.

Stub tower, only 4.5 m high.

Turbine shaft inclined at 30°.

One year warranty.

Ulrik Poulsen ApS,
Strandvejen 666,
2930 Klampenborg
Denmark
Telephone: (01) 63 65 66
Telex: 15764 Ulrik dk

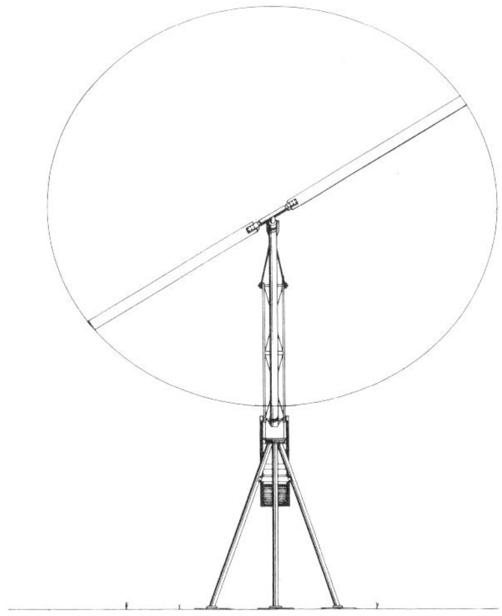

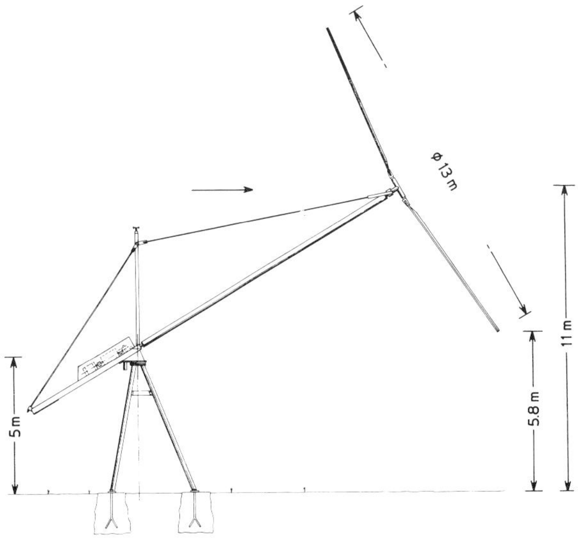

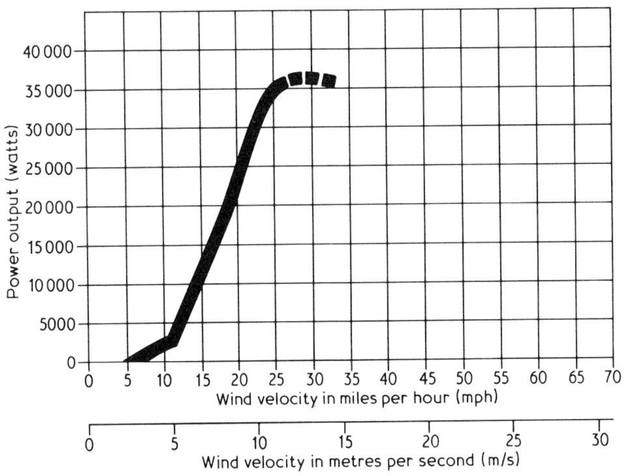

Hummingbird
A X P 6000

PERFORMANCE AND OPERATIONAL DATA

Rated electrical power	6000 watt
Rated output voltage	200 volt d.c.
Cut-in wind speed	3.0 m/s
Rated wind speed	11.6 m/s
Shut-down wind speed	27.7 m/s
Survival wind speed	55.5 m/s
Shaft speed at rated output	260 rev/min
Maximum shaft speed	400 rev/min

PHYSICAL DATA

Turbine

Type: Horizontal axis, upwind with tail vane
Dimensions
diameter 4.27 m

Blades

Construction: Epoxy resin with polyurethane foam core
Number: 3

Control: Fixed pitch turbine

Generator: Direct drive, brushless 3-phase synchronous a.c. generator with outer rotating p.m. field

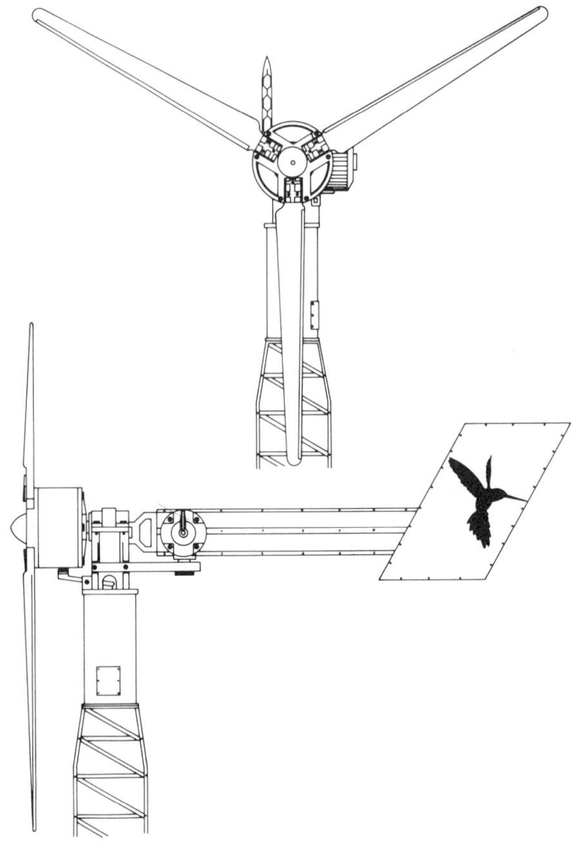

PRODUCTION

First production: 1979
Production quantity to date: over 100 (October 1982) including previous 4000/22 similar model

MISCELLANEOUS

One year warranty on parts and labour.
Annual inspection and 3 year regreasing schedule.

VARIANTS

UTI model operates through synchronous inverter for operation in parallel with 3 phase, 60 Hz network.
RIN model operates directly for battery charging in remote applications.

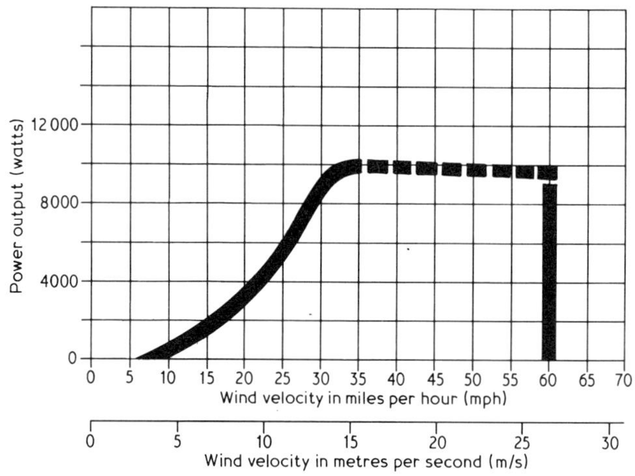

Power Group International Corp.,
13315 Stuebner-Airline Road, Suite 106,
Houston,
Texas 77014,
USA
Telephone: 713 444 5000
Telex: 79-2622

Sencenbaugh 24

PERFORMANCE AND OPERATIONAL DATA

Rated electrical power	24 watt
Rated output voltage	12 volt d.c.
Cut-in wind speed	4.5 m/s
Rated wind speed	9.4 m/s
Shut-down wind speed	none
Survival wind speed	44.6 m/s
Shaft speed at rated output	3200 rev/min
Maximum shaft speed	4500 rev/min

PHYSICAL DATA

Turbine
Type: Horizontal axis, upwind with tail vane
Dimensions
diameter 0.51 m
Blades
Construction: Cast aluminium
Number: 3
Control: Fixed pitch turbine
Generator: Permanent magnet a.c. generator (rectified)
Mechanical transmission: Direct coupling to turbine
System weight: 8.2 kg

PRODUCTION

First production: 1976
Production discontinued: January 1982

MISCELLANEOUS

One year warranty on parts and labour.
3 monthly, 6 monthly and annual maintenance items.
Handbook available.

Sencenbaugh Wind Electric,
253 Polaris Avenue,
Mountain View,
California,
USA
Telephone: (415) 964 1593

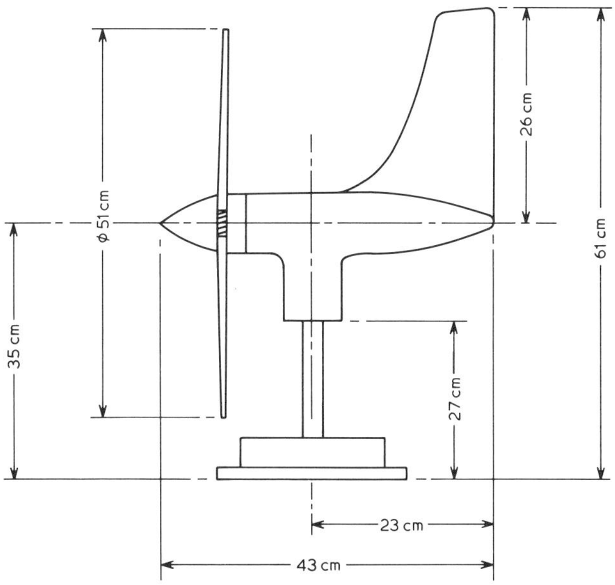

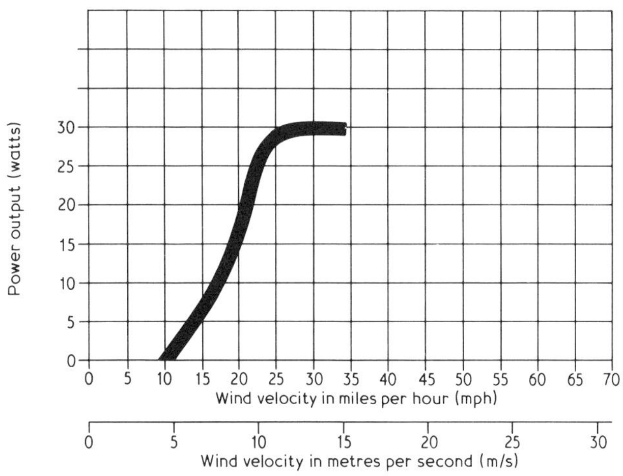

Sencenbaugh 500

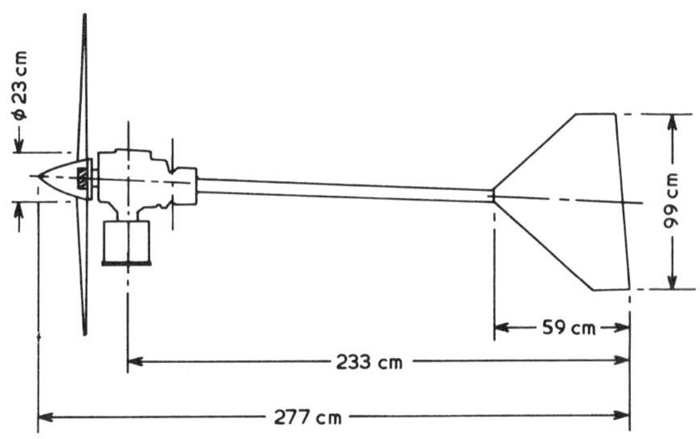

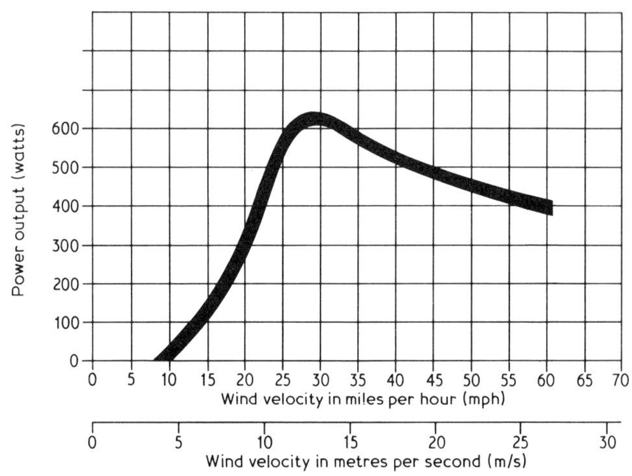

PERFORMANCE AND OPERATIONAL DATA

Rated electrical power	500 watt
Rated output voltage	12/24 volt d.c.
Cut-in wind speed	3.8 m/s
Rated wind speed	9.8 m/s
Shut-down wind speed	none
Survival wind speed	63.5 m/s
Shaft speed at rated output	1000 rev/min
Maximum shaft speed	1200 rev/min

PHYSICAL DATA

Turbine
Type: Horizontal axis, upwind with tail vane
Dimensions
diameter 2.03 m
Blades
Construction: Sitka spruce with polyurethane finish
Number: 3
Control: Fixed pitch turbine; high thrust turns turbine into plane of tail vane and wind
Generator: 12 pole, 3 phase a.c. generator
Mechanical transmission: Direct coupling to turbine
System weight
net 63.5 kg

PRODUCTION

First production: 1977
Production quantity to date: About 87 (September 1982)

MISCELLANEOUS

Two year warranty on parts and labour.
3 monthly, 6 monthly and annual maintenance items.
Handbook available.
Rohn towers model SSV (12.2 m or 18.3 m) or 25G (12.2 m, 15.3 m, 18.3 m, 21.4 m or 24.4 m) are offered.
Special features for hostile environments, including castings of corrosion resistant aluminium and weatherproof water-tight alternator sealing.

VARIANTS

12 volt option is known as 500–14.
24 volt option is known as 500–28.
48 volt option (500–48) is available standard in marine version with anodized castings and stainless steel hardware.

Sencenbaugh Wind Electric,
253 Polaris Avenue,
Mountain View,
California,
USA
Telephone: (415) 964 1593

Sencenbaugh 500-HDS

PERFORMANCE AND OPERATIONAL DATA

Rated electrical power	500 watt
Rated output voltage	12 volt d.c.
Cut-in wind speed	3.8 m/s
Rated wind speed	9.8 m/s
Shut-down wind speed	none
Survival wind speed	53.5 m/s
Shaft speed at rated output	1000 rev/min

PHYSICAL DATA

Turbine

Type: Horizontal axis, upwind with tail vane; hand operated winch tilts turbine into horizontal plane

Dimensions

diameter 2.03 m

Blades

Construction: Sitka spruce

Number: 3

Control: Fixed pitch turbine; high thrust tilts turbine up into horizontal plane

Generator: 12-pole a.c. generator

Mechanical transmission: Direct coupling to turbine

System weight

gross 43 kg

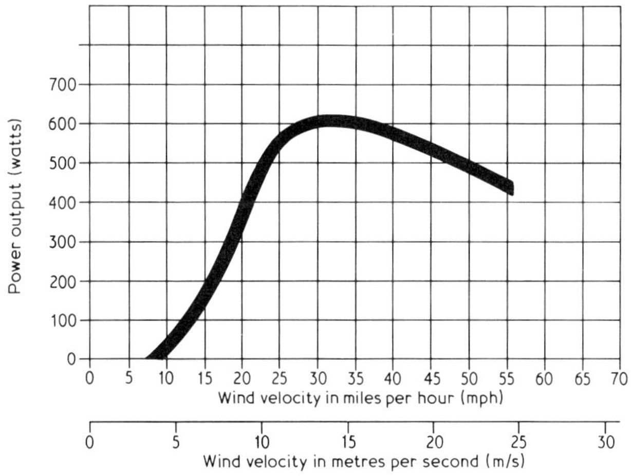

MISCELLANEOUS

One year warranty on parts and labour.

Suitable for Rohn guyed tower Model 25G (with KG-59 top) or SSV tower with special tower top section.

Sencenbaugh Wind Electric,
253 Polaris Avenue,
Mountain View,
California,
USA
Telephone: (415) 964 1593

Sencenbaugh 1000

PERFORMANCE AND OPERATIONAL DATA

Rated electrical power	1000 watt
Rated output voltage	12/24 volt d.c.
Cut-in wind speed	3.0 m/s
Rated wind speed	9.8 m/s
Shut-down wind speed	none
Survival wind speed	54 m/s
Shaft speed at rated output	350 rev/min
Maximum shaft speed	425 rev/min

PHYSICAL DATA

Turbine
Type: Horizontal axis, upwind with tail vane
Dimensions:
diameter 3.65 m

Blades
Construction: Sitka spruce with bonded copper leading edge and epoxy finish
Number: 3

Control: Fixed pitch turbine; high thrust turns turbine into plane of tail vane and wind

Generator: 3-phase, 12-pole a.c. generator (rectified)

Mechanical transmission: 3:1 ratio helical gearbox

System weight
net 81.7 kg

PRODUCTION

First production: 1976
Production quantity to date: 148 (September 1982)

MISCELLANEOUS

Two year warranty on parts and labour.
3 monthly, 6 monthly and annual maintenance items.
Handbook available.
Rohn tower model 25G (12.2 m, 15.3 m, 18.3 m, 21.4 m or 24.4 m) is offered.

VARIANTS

12 volt and 24 volt options are known as 1000-14 and 1000–28 respectively.
Marinized version available.

REPORTS

Rocky Flats Test Centre report:
Higashi, K.K. *Sencenbaugh Model 1000-14 Wind Turbine Generator, Performance Report,* RFP 3034/3533/79-5, July 1979.
Rocky Flats Performance Summary Sheet, Sencenbaugh Model 1000-14, August 1980.
Higashi, K.K. *Sencenbaugh Model 1000-14 Wind Turbine. Final Test Report,* RFP-3262, December 1980.

Sencenbaugh Wind Electric,
253 Polaris Avenue,
Mountain View,
California, USA
Telephone: (415) 964 1593

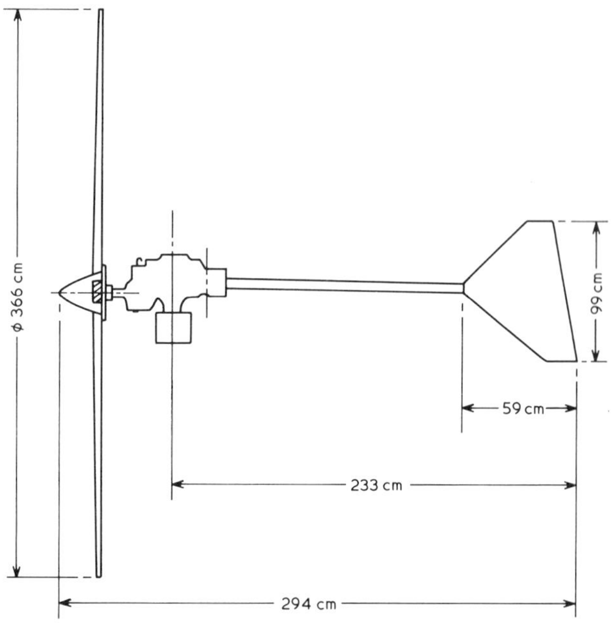

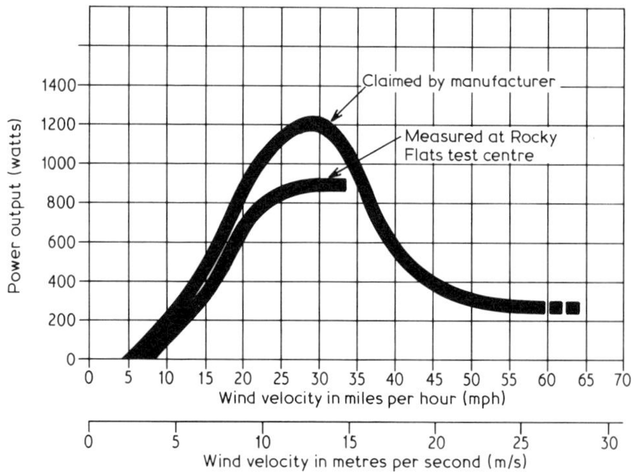

Sencenbaugh 1800

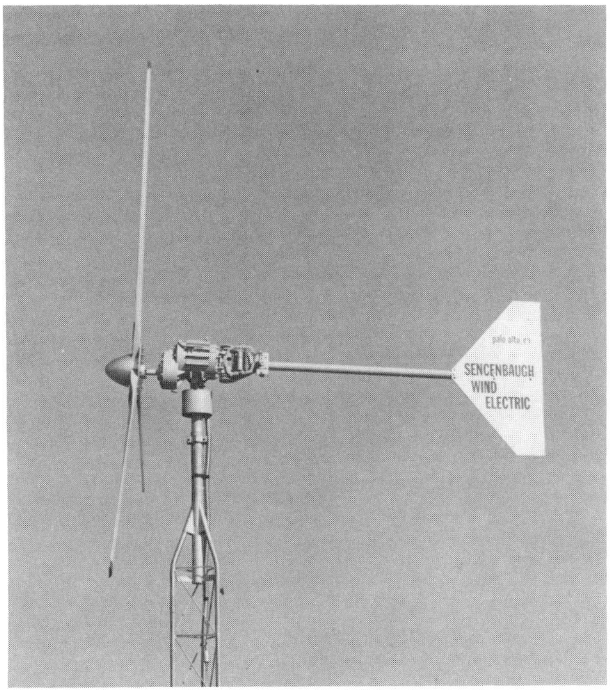

PERFORMANCE AND OPERATIONAL DATA

Rated electrical power	1800 watt
Rated output voltage	24 volt d.c.
Cut-in wind speed	3.0 m/s
Rated wind speed	11.6 m/s
Shut-down wind speed	none
Survival wind speed	54.0 m/s
Shaft speed at rated output	350 rev/min
Maximum shaft speed	425 rev/min

PHYSICAL DATA

Turbine
Type: Horizontal axis, upwind with tail vane
Dimensions
diameter 3.66 m
Blades
Construction: Sitka spruce with epoxy finish
Number: 3
Control: Fixed pitch turbine; high thrust turns turbine into plane of tail vane and wind
Generator: a.c. generator (rectified)
Mechanical transmission: 4:1 ratio helical gearbox
System weight
net 86 kg

PRODUCTION

First production: 1977
Production quantity to date: 26 (September 1982)

MISCELLANEOUS

Two year warranty on parts and labour.
3 monthly, 6 monthly and annual maintenance items.
Handbook available.
Rohn tower model 25G (12.2 m, 15.3 m, 18.3 m, 21.4 m or 24.4 m) is offered.
Produced on special order.

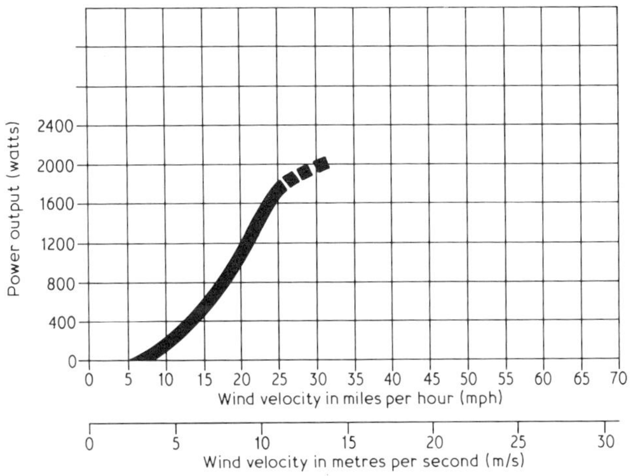

Sencenbaugh Wind Electric,
253 Polaris Avenue,
Mountain View,
California,
USA
Telephone: (415) 964 1593

SJ 10 kW-R

PERFORMANCE AND OPERATIONAL DATA

Rated electrical power	10 000 watt
Rated output voltage	380 volt a.c.
Cut-in wind speed	5.0 m/s
Rated wind speed	12.0 m/s
Shut-down wind speed	above 12.0 m/s
Survival wind speed	25 m/s for a maximum of 10 s
Shaft speed at rated output	50 rev/min
Maximum shaft speed	95 rev/min (at 25 m/s for a maximum of 10 s)

PHYSICAL DATA

Turbine
Type: Horizontal axis, upwind with tail vane
Dimensions
diameter 6.30 m
Blades
Construction: Polyurethane
Number: 16
Control: Fixed pitch turbine
Generator: Synchronous a.c. generator
Mechanical transmission: Speed increasing 'UniMax' toothed gear drive
System weight
net 1150 kg

PRODUCTION

First production: 1978
Production quantity to date: Over 350 (1980)

MISCELLANEOUS

One year warranty on complete installation.
Maintenance comprises six monthly hydraulics inspection, annual gearbox oil change, bi-annual change of hydraulic oil.
Operates at variable voltage and frequency, suitable for heating purposes.
Tower height 14.0 m.

VARIANTS

Equipment for constant voltage and equipment for water pumping can also be delivered.

REPORTS

Rasmussen, F. *Effectmalinger pa SJ-10kW-R Vindmolle* Risø National Laboratory, Roskilde, Denmark, March 1980.

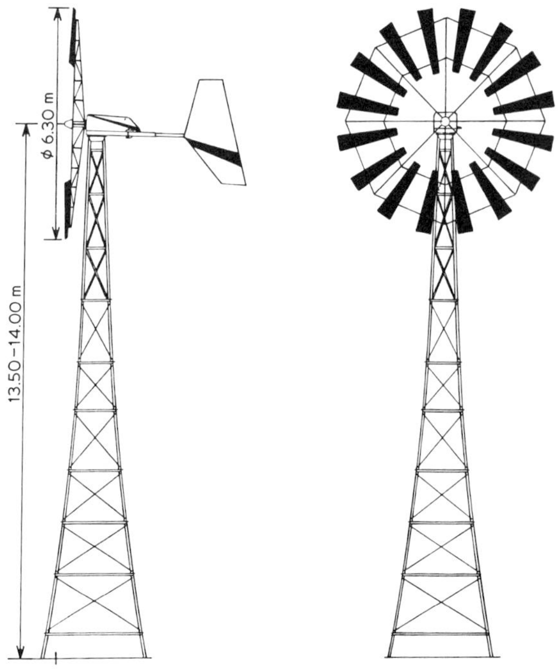

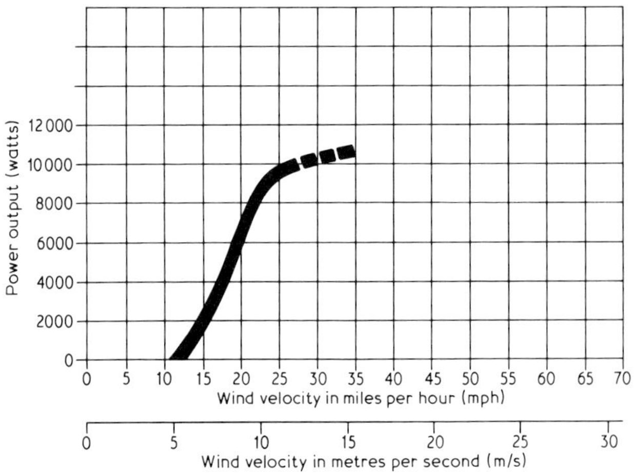

SJ Wind Power ApS,
Suderbovej 4,
DK 9900 Frederikshavn,
Denmark
Telephone: (08) 43 00 33

Wesco A30

PERFORMANCE AND OPERATIONAL DATA

Rated electrical power	2200 watt
Cut-in wind speed	3.5–5.0 m/s
Rated wind speed	10.0 m/s
Survival wind speed	55 m/s
Maximum shaft speed	400 rev/min

PHYSICAL DATA

Turbine

Type: Horizontal axis, upwind with tail vane

Dimensions

diameter 3.0 m

Blades

Number: 3

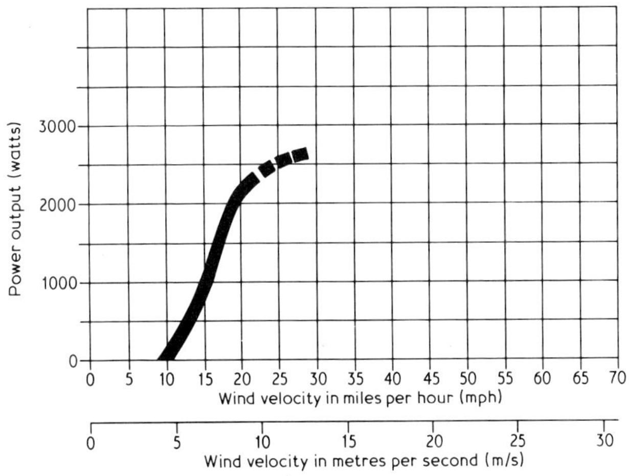

Sunwind Energy,
Route de l'Abbaye 5,
Ch-2022 Bevaix,
Switzerland
Telephone: 036 46 1404
Telex: 35100

Tornado
5 W

PERFORMANCE AND OPERATIONAL DATA

Rated electrical power	5 watt
Rated output voltage	12 volt d.c.
Cut-in wind speed	4.3 m/s
Rated wind speed	9.8 m/s
Shut-down wind speed	none

PHYSICAL DATA

Turbine
Type: Horizontal axis, upwind with tail vane
Dimensions
diameter 0.56 m
Blades
Construction: Pressed aluminium sheet
Number: 6
Control: Fixed pitch turbine
Generator: Sturmey Archer (bicycle type) alternator
Mechanical transmission: Direct coupling to turbine
System weight
net 2.7 kg

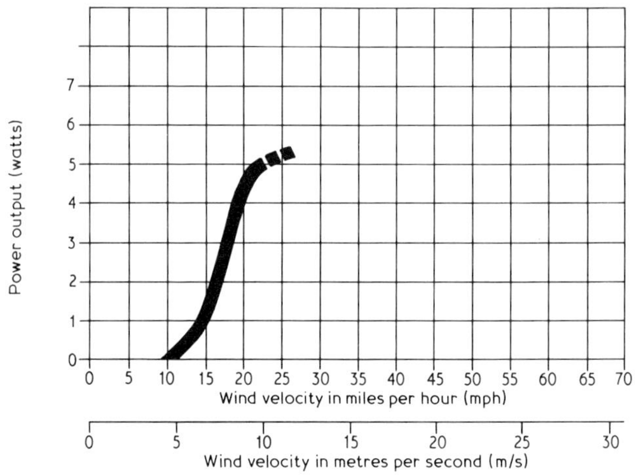

Tornado Wind Generators,
75 Benslow Lane,
Hitchin,
Herts SG4 9RA
UK
Telephone: (0462) 59457
Telex: 897836

Tornado 500 W

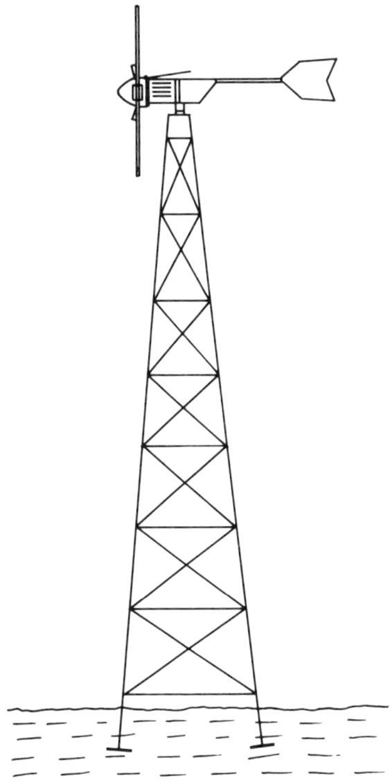

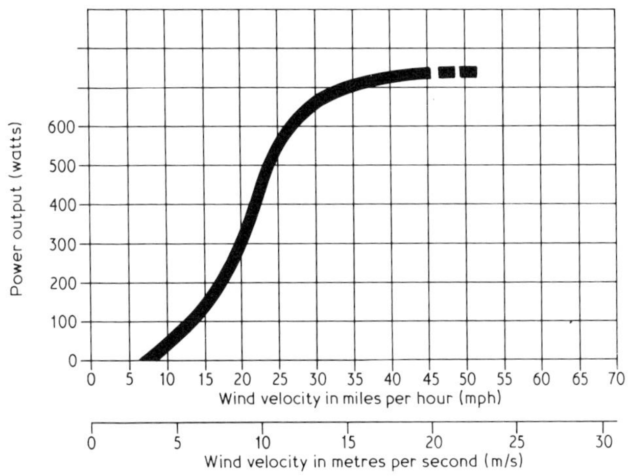

PERFORMANCE AND OPERATIONAL DATA

Rated electrical power	500 watt
Rated output voltage	12/24/36/110 volt d.c.
Cut-in wind speed	3.5 m/s
Rated wind speed	11.0 m/s
Shut-down wind speed (manual)	~23 m/s
Survival wind speed	45 m/s
Shaft speed at rated output	700 rev/min
Maximum shaft speed	1500 rev/min

PHYSICAL DATA

Turbine

Type: Horizontal axis, upwind with tail vane

Dimensions

diameter 2.1 m

Blades

Construction: Epoxy impregnated wood with plastic leading edge protection

Number: 2 (main); 8+spoiler (starting)

Control: Fixed pitch turbine

Generator: Brushless a.c. generator (rectified)

Mechanical transmission: Direct coupling to turbine

System weight

net 50 kg

PRODUCTION

First production: Spring 1980

Production quantity to date: 57 (August 1982)

MISCELLANEOUS

4 m, 8 m or 10 m towers available.

3 monthly visual check and slip ring lubrication, 12 monthly visual check, grease bearings, slip ring lubrication and paintwork touch-up.

One year guarantee on wind turbine unit, two year guarantee on alternator.

VARIANTS

Automatic shut-down facility can be added.

Water/space heating facility can be added.

Tornado Wind Generators,
75 Benslow Lane,
Hitchin,
Herts SG4 9RA,
UK
Telephone: (0462) 59457
Telex: 897836

Trimblemill

PERFORMANCE AND OPERATIONAL DATA

Rated electrical power 5000 watt
Rated output voltage 220 volt a.c.
Cut-in wind speed 4–5 m/s
Rated wind speed 10 m/s
Shut-down wind speed 20–25 m/s (although partial generation is maintained above this wind speed)
Survival wind speed 45 m/s
Maximum shaft speed 250 rev/min

PHYSICAL DATA

Turbine

Type: Horizontal axis, upwind with tail vane; two contrarotating turbines
Dimensions
diameter 6 m

Blades

Construction: GRP envelope separated by high density polyurethane foam mounted on a rigid beam
Number: 3 (front); 5 (rear)
Control: Fixed pitch blades, but can be adjusted
Generator: Multi-pole contrarotating a.c. generator
Mechanical transmission: Direct coupling to turbine
System weight: 635 kg

PRODUCTION

First production: Early 1980
Production quantity to date: 17 (September 1982)

MISCELLANEOUS

9.3 m tower designed for ease of erection in remote areas and ease of lowering for inspection or blade changing.
Complete 'packages' available for storing energy as either hot water, hot air or in batteries.

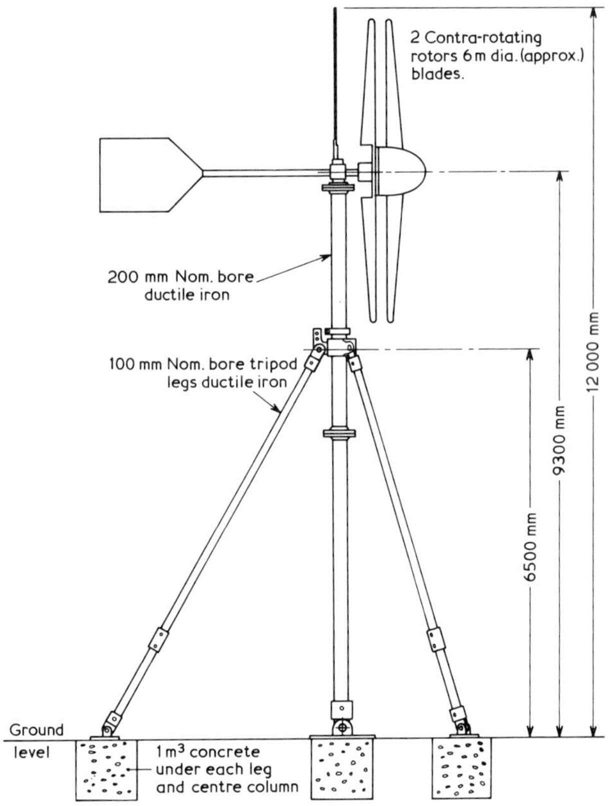

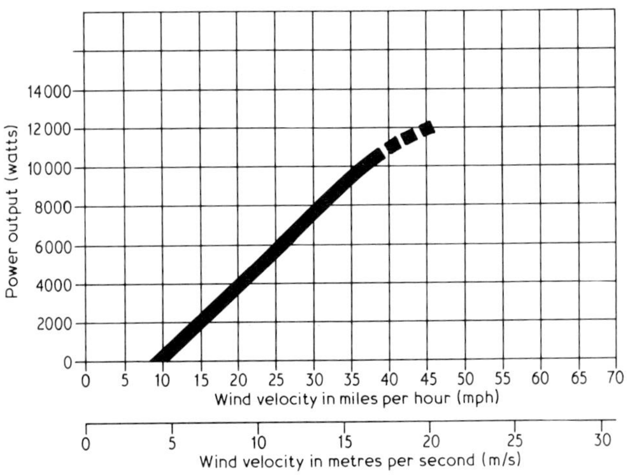

Trimble Windmills,
c/o NEI Clarke Chapman Ltd,
Victoria Works,
Gateshead,
Tyne and Wear NE8 3HS
UK
Telephone: (0423) 56006

Wind Titan 500 watt

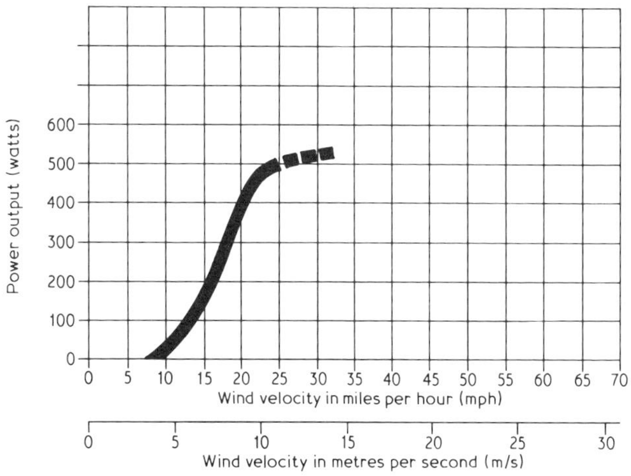

PERFORMANCE AND OPERATIONAL DATA

Rated electrical power	500 watt
Rated output voltage	12 volt d.c.
Cut-in wind speed	about 3.8 m/s
Rated wind speed	11.1 m/s
Shut-down wind speed	13.4 m/s to 26.8 m/s (adjustable)
Survival wind speed	49.1 m/s
Maximum shaft speed	234 rev/min

PHYSICAL DATA

Turbine
Type: Horizontal axis, upwind with tail vane
Dimensions
diameter 3.05 m
Blades
Construction: Aluminium or spruce
Number: 2
Control: Fixed pitch turbine
Generator: a.c. generator
Mechanical transmission: 15:1 ratio speed increasing gear and belt drive
System weight
net 88 kg

PRODUCTION

First production: 1975
Production quantity to date: About 8 (November 1979)

MISCELLANEOUS

All parts are warrantied against defects under normal operation for a period of one year.

TWR Enterprises,
72 West Meadow Lane,
Sandy,
Utah 84070,
USA

Wind Titan 1200 watt

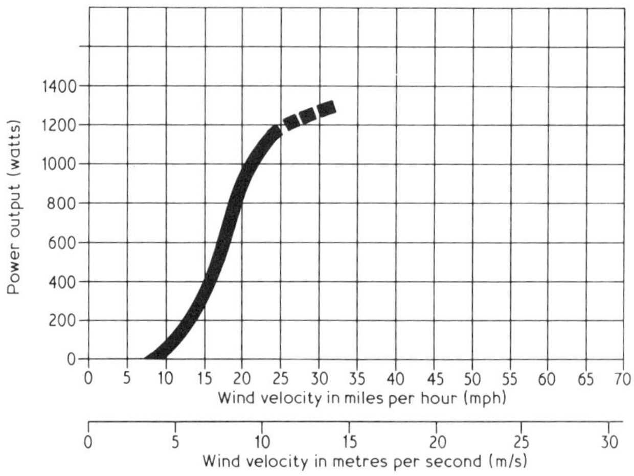

PERFORMANCE AND OPERATIONAL DATA

Rated electrical power	1200 watt
Rated output voltage	12/24 volt d.c.
Cut-in wind speed	about 3.3 m/s
Rated wind speed	11.1 m/s
Shut-down wind speed	13.4–26.8 m/s (adjustable)
Survival wind speed	49.1 m/s
Maximum shaft speed	234 rev/min

PHYSICAL DATA

Turbine
Type: Horizontal axis, upwind with tail vane
Dimensions
diameter 4.27 m
Blades
Construction: Aluminium or spruce
Number: 2
Control: Variable pitch turbine
Generator: a.c. generator
Mechanical transmission: 15:1 ratio speed increasing gear and belt drive
System weight
net 114 kg

PRODUCTION

First production: 1975
Production quantity to date: About 13 (November 1979)

MISCELLANEOUS

All parts are warrantied against defects under normal operation for a period of one year.

TWR Enterprises,
72 West Meadow Lane,
Sandy,
Utah 84070,
USA

Wind Titan 3000 watt

PERFORMANCE AND OPERATIONAL DATA

Rated electrical power	3000 watt
Rated output voltage	32/36 volt d.c.
Cut-in wind speed	about 3.5 m/s
Rated wind speed	11.1 m/s
Shut-down wind speed	13.4–26.8 m/s (adjustable)
Survival wind speed	49.1 m/s
Maximum shaft speed	180 rev/min

PHYSICAL DATA

Turbine
Type: Horizontal axis, upwind with tail vane
Dimensions
diameter 5.50 m
Blades
Construction: Aluminium or spruce
Number: 3
Control: Variable pitch turbine
Generator: a.c. generator
Mechanical transmission: 20:1 ratio speed increasing gear and belt drive
System weight
net 204 kg

PRODUCTION

First production: 1975
Production quantity to date: About 5 (November 1979)

MISCELLANEOUS

All parts are warrantied against defects under normal operation for a period of one year.

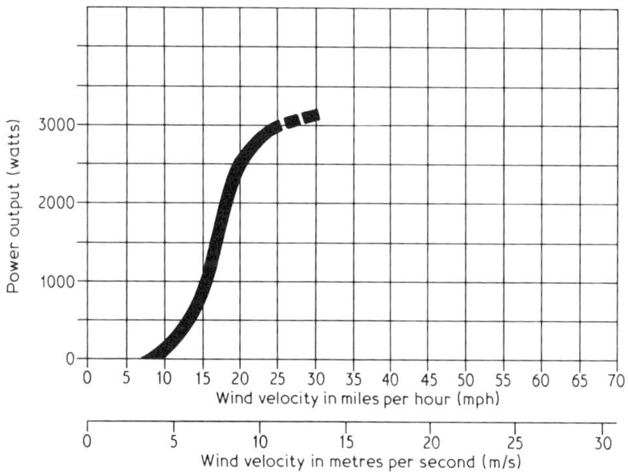

TWR Enterprises,
72 West Meadow Lane,
Sandy,
Utah 84070,
USA

Vendelbo-Mølle Type 10

PERFORMANCE AND OPERATIONAL DATA

Rated electrical power	10000 watt
Rated output voltage	380/220 volt
Maximum shaft speed	75 rev/min

PHYSICAL DATA

Turbine
Type: Horizontal axis, upwind with tail vane
Dimensions
diameter 5.5 m
Blades
Construction: Steel with 'polypaste'
Number: 5
Control: Centrifugally operated blade pitch control
Mechanical transmission: 20:1 ratio speed increasing transmission

PRODUCTION

Production quantity to date: 3 (July 1980)

MISCELLANEOUS

Tower of angle iron, offered in various heights, 10 m and 12 m standard lengths.
Designed for heating application.
Has been passed by RISØ proving station as suitable for subsidy in Denmark.
Type 15 (7.0 m diameter turbine, rated power 15000 watt) under development.

Vendelbo Trapper,
Gølstrup Hede 93,
9480 Løkken,
Denmark
Telephone: (08) 996244

Vestas 30

PERFORMANCE AND OPERATIONAL DATA

Rated electrical power	30 000 watt (primary) 5500 watt (secondary)
Rated output voltage in parallel with 3 phase, 50 Hz network)	380 volt (operated
Cut-in wind speed	4 m/s
Rated wind speed	14 m/s
Shut-down wind speed	25–30 m/s
Survival wind speed	~ 50 m/s (probably)
Shaft speed at rated output	83 rev/min

PHYSICAL DATA

Turbine

Type: Horizontal axis

Dimensions

diameter 10 m

Blades

Construction: Reinforced polyester

Number: 3

Control: Fixed pitch turbine with built-in air brakes of stainless steel centrifugally released mechanism

Generator: 30 kW induction generator and 5 kW induction generator for efficient production in low wind

Mechanical transmission: Speed increasing gearbox

System weight: 5900 kg

PRODUCTION

First production: January 1980

Production quantity to date: 23 (September 1982)

MISCELLANEOUS

Annual service check.
18 m or 24 m high lattice tower.
12 month warranty.
Not for export until end 1981.
Production taken over from Herborg Vindkraft.

VARIANTS

Rated electrical power 22 000 watt (primary); 3000 watt (secondary).

Associated with lower rated power, shaft speed at rated power 75 rev/min.

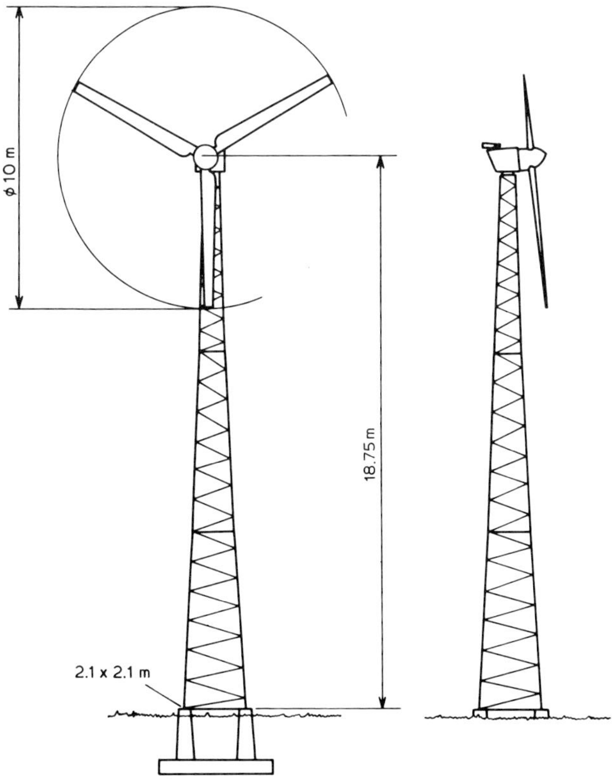

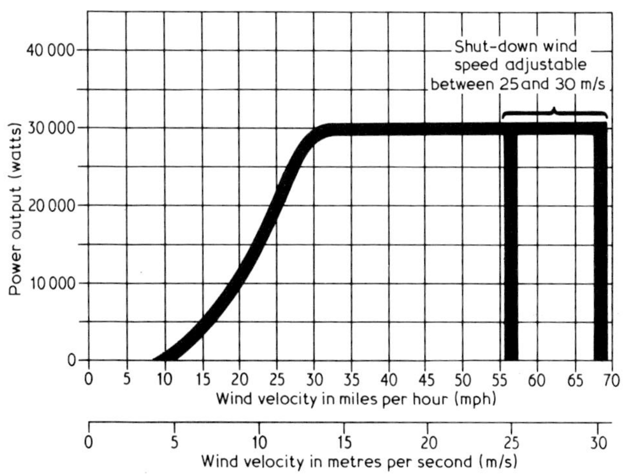

A/S Vestas,
Jernbangade,
DK 6940 Lem,
Denmark
Telephone: 07-341188
Telex: 60733

Vestas 55

PERFORMANCE AND OPERATIONAL DATA

Rated electrical power 55 000 watt (primary) 11 000 watt (secondary)
Rated output voltage 380 volt (operated in parallel with 3-phase 50 Hz network)
Cut-in wind speed 4 m/s
Rated wind speed 12.5 m/s
Shut-down wind speed 25–30 m/s
Survival wind speed ~ 50 m/s (probably)
Shaft speed at rated output 52 rev/min

PHYSICAL DATA

Turbine
Type: Horizontal axis
Dimensions
diameter 15 m
Blades
Construction: Reinforced polyester
Number: 3
Control: Fixed pitch turbine with built-in air brakes of stainless steel centrifugally-released mechanism
Generator: 55 kW induction generator and 11 kW induction generator for efficient production in low wind
Mechanical transmission: Speed-increasing gearbox
System weight: 9300 kg

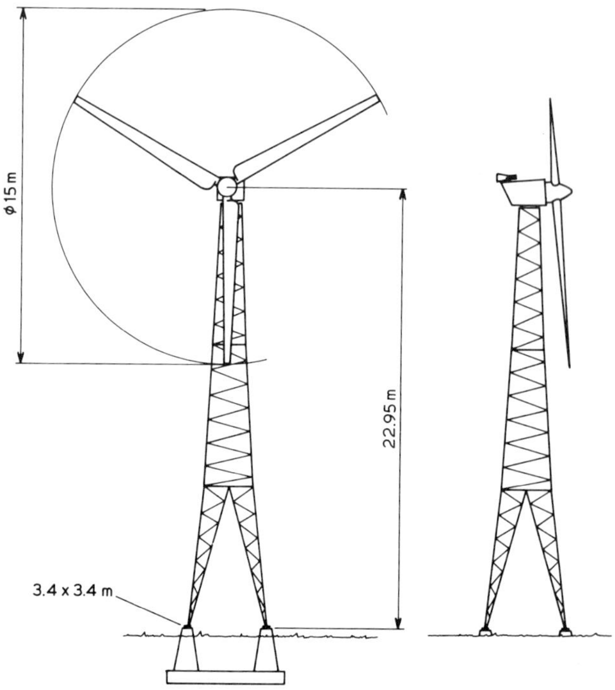

PRODUCTION

First production: January 1980
Production quantity to date: 61 (September 1982)

MISCELLANEOUS

Annual service check.
18 m or 22 m high lattice tower.
12 month warranty.
Not for export until end 1982.

VARIANTS

Rated electrical power 45 000 watt (primary); 7500 watt (secondary).

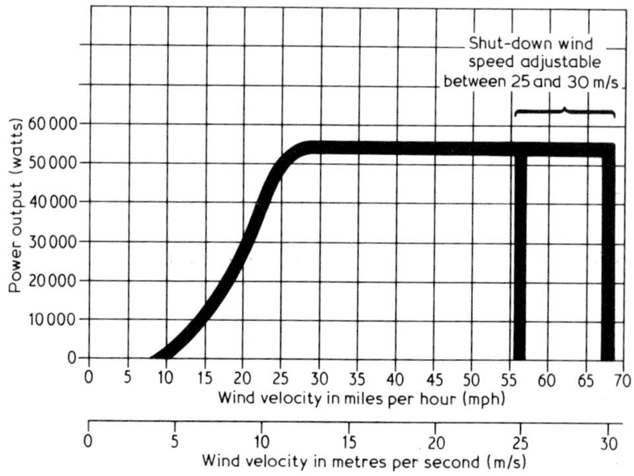

A/S Vestas,
Jernbangade,
DK 6940 Lem,
Denmark
Telephone: 07-341188
Telex: 60733

Vølund 15 kW

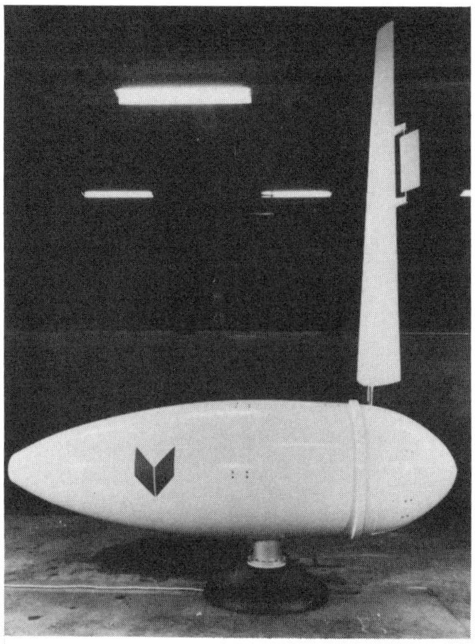

PERFORMANCE AND OPERATIONAL DATA

Rated electrical power	15 000 watt
Rated output voltage	380 volt a.c.
(operated in parallel with 3-phase, 50 Hz network)	
Cut-in wind speed	5 m/s
Rated wind speed	11 m/s
Shut-down wind speed	30 m/s
Survival wind speed	70 m/s
Shaft speed at rated output	120 rev/min
Maximum shaft speed	130 rev/min

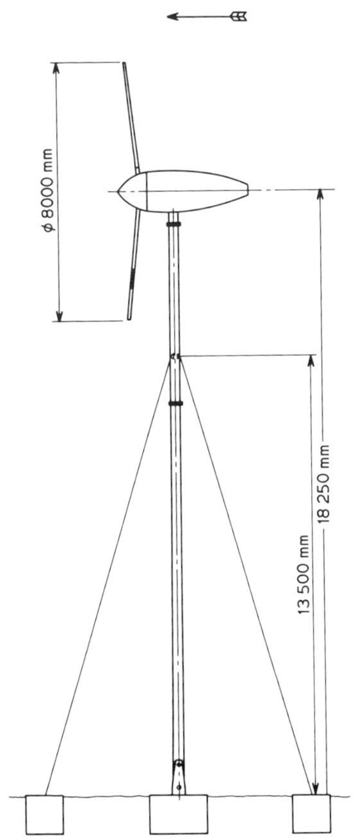

PHYSICAL DATA

Turbine

Type: Horizontal axis downwind

Dimensions

diameter 8.0 m

Blades

Construction: Fibreglass reinforced polyester, shell construction

Number: 2

Control: Fixed pitch turbine with flaps

Generator: Induction generator

Mechanical transmission: Speed-increasing gearbox

System weight: 2500 kg (including tower)

PRODUCTION

First production: August 1981

Production quantity to date: 5 (June 1981)

MISCELLANEOUS

Offered complete with 18 m tubular guyed steel tower.
One year guarantee.
Half-yearly maintenance.

VARIANTS

Single phase 380 volt a.c. machine with direct coupled 42 pole permanent magnet synchronous a.c. generator, cut-in wind speed 3 m/s, rated shaft speed 140 rev/min.

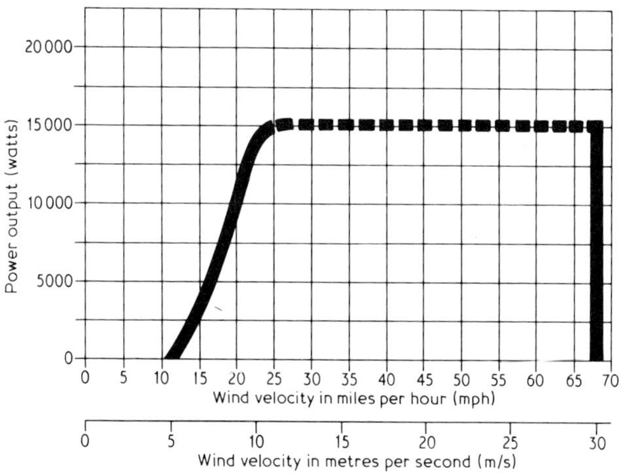

Vølund Fibreglass and Windtechnic A/S,
Marsk Stigsvej 4,
DK-8800 Viborg,
Denmark
Telephone: 456623499
Telex: 66225

Whirlwind Power Co
Model A120

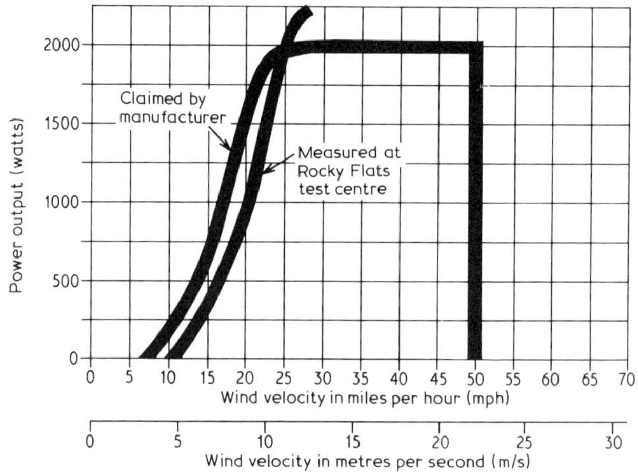

PERFORMANCE AND OPERATIONAL DATA

Rated electrical power	2000 watt
Rated output voltage	90–110 volt a.c. or 120 volt d.c.
Cut-in wind speed	3.1 m/s
Rated wind speed	11.2 m/s
Shut-down wind speed	22.3 m/s
Survival wind speed	35.7 m/s
Shaft speed at rated output	900 rev/min
Maximum shaft speed	2000 rev/min

PHYSICAL DATA

Turbine
Type: Horizontal axis, downwind, operating freely in yaw
Dimensions
diameter 3.05 m
Blades
Construction: Sitka spruce with polyurethane finish
Number: 2
Control: Fixed pitch turbine
Generator: 3-phase, permanent magnet, 18 pole a.c. generator
Mechanical transmission: Direct coupling to turbine
System weight
net 32 kg

PRODUCTION

First production: 1978

MISCELLANEOUS

One year warranty on parts and labour.
Maintenance every five years; lubricate bearings, refinish blades.

REPORTS

Rocky Flats Performance Summary Sheet Whirlwind A120, December 1980.

Whirlwind Power Company,
2458 W. 29th Avenue,
Denver,
Colorado 80211,
USA
Telephone: (303) 477 6436

Winco
Wincharger W200

PERFORMANCE AND OPERATIONAL DATA

Rated electrical power	200 watt
Rated output voltage	12/24 volt d.c.
Cut-in wind speed	4.0 m/s
Rated wind speed	10.3 m/s
Shut-down wind speed	none
Survival wind speed	31.2 m/s
Shaft speed at rated output	900 rev/min
Maximum shaft speed	900 rev/min

PHYSICAL DATA

Turbine
Type: Horizontal axis, upwind with tail vane
Dimensions
diameter 1.83 m
Blades
Construction: Varnished redwood with polyurethane film leading edge
Number: 2
Control: Fixed pitch blades with centrifugally operated air brake governor
Generator: Self-excited d.c. generator
Mechanical transmission: Direct coupling to turbine
System weight
net 27 kg

PRODUCTION

First production: 1936
Production quantity to date: 1000s

MISCELLANEOUS

3.1 m, 4 leg angle iron tower offered as standard.
One year warranty on parts and labour.
Annual maintenance comprising oiling and inspection.
The 12 and 24 volt options are coded W200-12 and W200-24 respectively. Cut-in wind speed is 4.5 m/s for the W200-24.

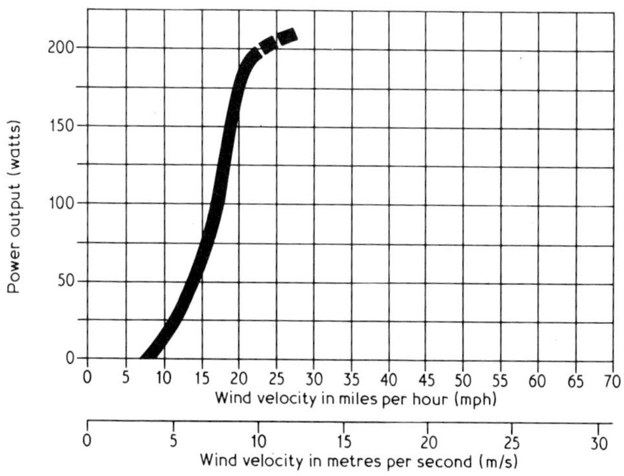

Winco-Division of Dyna Technology,
7850 Metro Parkway,
Minneapolis,
Minnesota 55420,
USA
Telephone: (612) 853 8400

Winco Wincharger W450

PERFORMANCE AND OPERATIONAL DATA

Rated electrical power	450 watt
Rated output voltage	12/24/36/48 volt d.c.
Cut-in wind speed	4.9 m/s
Rated wind speed	11.2 m/s
Shut-down wind speed	none

PHYSICAL DATA

Turbine
Type: Horizontal axis, upwind turbine with tail vane
Dimensions
diameter 2.44 m
Blades
Construction: Close grain laminated hardwood with polyurethane film leading edge
Number: 2
Control: Fixed pitch blades with centrifugally operated air brake governor
Generator: Self-excited d.c. generator
Mechanical transmission: Direct coupling to turbine
System weight: 45 kg

MISCELLANEOUS

3.1 m, 4 leg angle iron tower offered as standard.
One year warranty on parts and labour.
Annual maintenance, comprising oiling and inspection.
The 12, 24, 36 and 48 volt options are coded W450-12, W450-24, W450-36 and W450-48 respectively.

Winco-Division of Dyna Technology,
7850 Metro Parkway,
Minneapolis,
Minnesota 55420
USA
Telephone: (612) 853 8400

Wingen 25

PERFORMANCE AND OPERATIONAL DATA

Rated electrical power	25 000 watt
Rated output voltage	230/460 volt
Cut-in wind speed	4.5–5.4 m/s
Rated wind speed	11.2 m/s
Shut-down wind speed	20.0 m/s
Survival wind speed	44.6 m/s
Shaft speed at rated output	67 rev/min
Maximum shaft speed	~ 80 rev/min

PHYSICAL DATA

Turbine
Type: Horizontal axis, downwind type, operating freely in yaw
Dimensions
diameter 13.1 m
Blades
Construction: Aluminium with spar, ribs and skin of fibreglass/silicone
Number: 3
Control: Variable pitch turbine with microprocessor control
Generator: 4 pole, 1800 rev/min induction generator at ground level
Mechanical transmission: Right-angle speed increasing gearbox (approx. 25:1 ratio) housed in nacelle
System weight
1135 kg (turbine)
680 kg (tower)

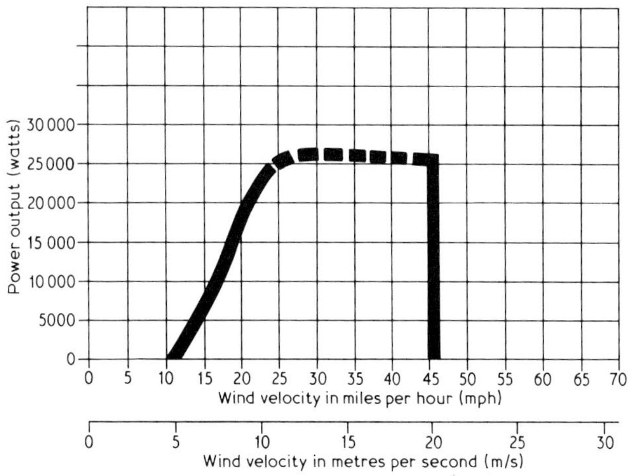

PRODUCTION

First production: January 1980
Production quantity to date: 8 (July 1981)

MISCELLANEOUS

One year full warranty.
Annual maintenance requirement.

VARIANTS

Mechanical coupling to a variety of loads.
19.65 m (standard), 26.2 m or 32.75 m high free standing towers offered.
Operating characteristic can be changed through software modifications.
New model rated 27 000 watt at 11.2 m/s under development.

Wind Engineering Corp,
Airport Industrial Area,
PO Box 5936,
Lubbock,
Texas 79417,
USA
Telephone: (806) 763 3182

Windfos 22

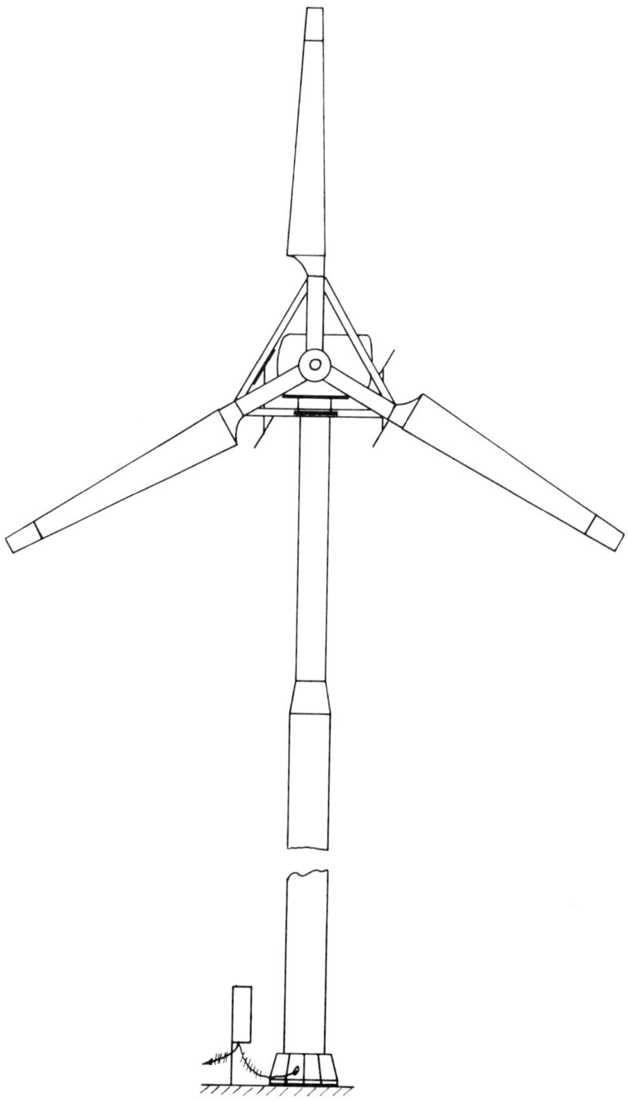

PERFORMANCE AND OPERATIONAL DATA

Rated electrical power	22 000 watt
Rated output voltage	380/220 volt a.c.
(operated in parallel with 3-phase, 50 Hz network)	
Cut-in wind speed	4 m/s
Rated wind speed	12–13 m/s
Shut-down wind speed	25 m/s
Survival wind speed	40–60 m/s (approx.)
Shaft speed at rated output	85 rev/min

PHYSICAL DATA

Turbine

Type: Horizontal axis, upwind with twin fan tails
Dimensions
diameter 10.6 m

Blades

Construction: Polyester/glassfibre (both skin and structure) with rotating tip air brakes
Number: 3

Control: Fixed pitch turbine; rotating tip air brakes on blades

Generator: Induction generator

Mechanical transmission: Speed increasing helical gearbox

PRODUCTION

First production: 1980
Production quantity to date: 2 (1980)

MISCELLANEOUS

One or three years warranty offered.
Annual maintenance comprises greasing, gearbox oil change every 5 years.
Different heights of tower are offered.

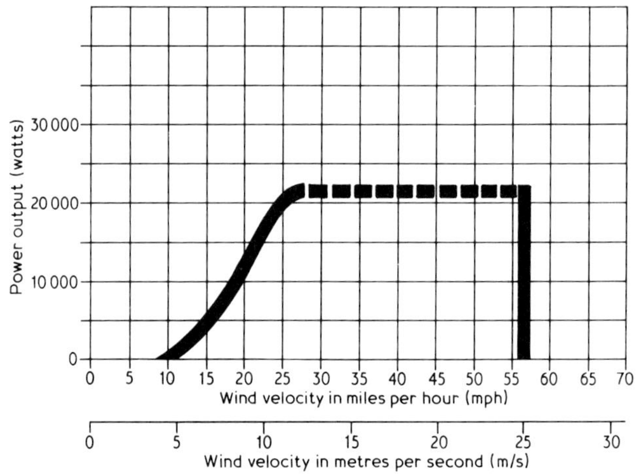

Windfos,
Nordvestvej 4,
9000 Aalborg,
Denmark
Telephone: (08) 13 92 15

Windfos 30

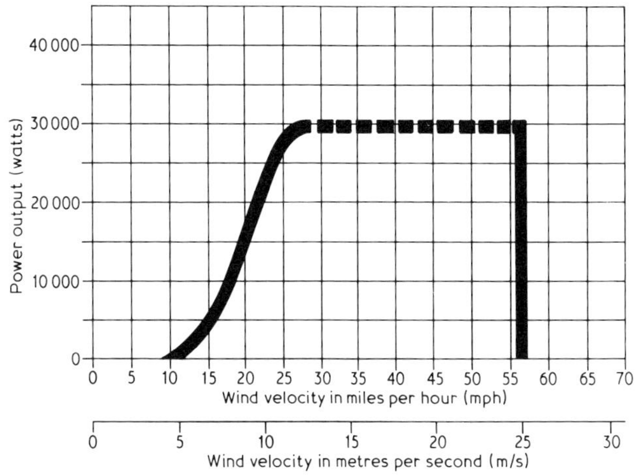

PERFORMANCE AND OPERATIONAL DATA

Rated electrical power	30 000 watt
Rated output voltage	380/220 volt a.c.
(operated in parallel with 3-phase, 50 Hz network)	
Cut-in wind speed	4 m/s
Rated wind speed	12–13 m/s
Shut-down wind speed	25 m/s
Survival wind speed	40–60 m/s (approx.)
Shaft speed at rated output	79 rev/min

PHYSICAL DATA

Turbine
Type: Horizontal axis, upwind with twin fan tails
Dimensions
diameter 13.2 m
Blades
Construction: Polyester fibre (both skin and structure) with rotating tip air brakes
Number: 3
Control: Fixed pitch turbine
Generator: Induction generator
Mechanical transmission: Speed increasing helical gearbox

PRODUCTION

First production: 1980
Product quantity to date: 2 (1980)

MISCELLANEOUS

One or three year warranty offered.
Annual maintenance comprises greasing, gearbox oil change every 5 years.
Different heights of tower are offered.

Windfos,
Nordvestvej 4,
9000 Aalborg,
Denmark
Telephone: (08) 13 92 15

Wind-Matic WM 10S

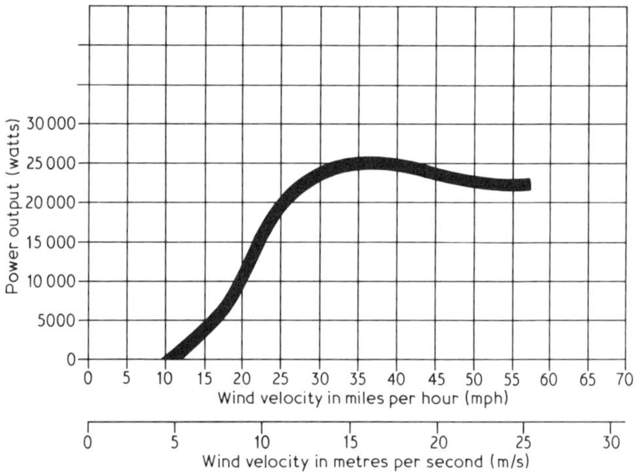

PERFORMANCE AND OPERATIONAL DATA

Rated electrical power 22 000 watt
Rated output voltage 380 volt a.c.
(operated in parallel with 3-phase 50 Hz network)
Cut-in wind speed 4.5 m/s
Rated wind speed 12 m/s
Maximum shaft speed 69 rev/min

PHYSICAL DATA

Turbine
Type: Horizontal axis, upwind with twin fan tails
Dimensions
diameter 10.2 m
Blades
Construction: Fibreglass with spar of laminated wood
Number: 3
Control: Fixed pitch turbine with flaps
Generator: Induction generator
Mechanical transmission: Speed increasing gearbox

PRODUCTION

First production: 1979
Production quantity to date: Over 65 in total (all types) (July 1981)

MISCELLANEOUS

Offered complete with 18 m or 22.5 m, 4 leg, lattice tower prefabricated in two sections.
Previously produced and sold under the name 'Riisager-Mollen'.
Wind-Matic is affiliated to the HOLEC-Group.

Wind-Matic A/S,
Industrivej Nord 15,
Birk,
DK-7400 Herning,
Denmark
Telephone: (07) 127700
Telex: 62250

Wind-Matic WM 12S

PERFORMANCE AND OPERATIONAL DATA

Rated electrical power	30 000 watt
Rated output voltage	380 volt a.c.
(operated in parallel with 3-phase, 50 Hz network)	
Cut-in wind speed	4.5 m/s
Rated wind speed	12 m/s
Maximum shaft speed	54 rev/min

PHYSICAL DATA

Turbine
Type: Horizontal axis, upwind with twin fan tails
Dimensions
diameter 12.5 m

Blades
Construction: Fibreglass with spar of laminated wood
Number: 3

Control: Fixed pitch turbine, with flaps
Generator: Induction generator
Mechanical transmission: Speed increasing gearbox

PRODUCTION

First production: 1979
Production quantity to date: Over 65 produced in total (all types) (July 1981)

MISCELLANEOUS

Offered complete with 18 m or 22.5 m, 4 leg, lattice tower, prefabricated in three main sections.
Previously produced and sold under the name 'Riisager-Mollen'.
Wind-Matic is affiliated to the HOLEC-Group.

REPORTS

Rasmussen, F. *Effektmalinger pa 30 kW Riisagermolle* Risø National Lab, Roskilde, Denmark, Report No RISO-M-2227, March 1980.

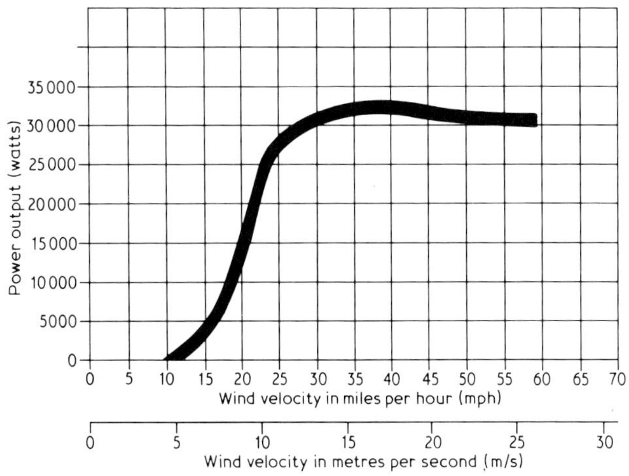

Wind-Matic A/S,
Industrivej Nord 15,
Birk,
DK-7400 Herning,
Denmark
Telephone: (07) 127700
Telex: 62250

Wind-Matic WM 14S

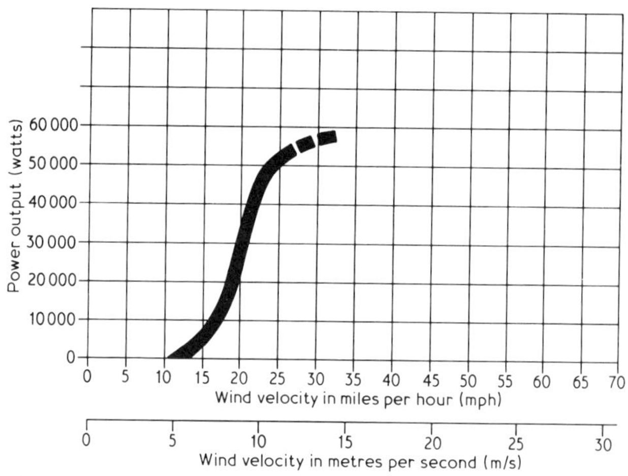

PERFORMANCE AND OPERATIONAL DATA

Rated electrical power 55 000 watt
Rated output voltage 380 volt a.c.
(operated in parallel with 3-phase, 50 Hz network)
Cut-in wind speed 4.5 m/s
Rated wind speed 12 m/s
Maximum shaft speed 50 rev/min

PHYSICAL DATA

Turbine
Type: Horizontal axis, upwind with twin fan tails
Dimensions
diameter 14.5 m
Blades
Construction: Fibreglass with spar of laminated wood
Number: 3
Control: Fixed pitch turbine, with flaps
Generator: Induction generator
Mechanical transmission: 20:1 ratio speed-increasing gearbox

PRODUCTION

First production: 1979
Production quantity to date: Over 65 in total (all types) (July 1981)

MISCELLANEOUS

Offered complete with 18 m or 22.5 m, 4 leg, lattice tower, prefabricated in three main sections.
Previously produced and sold under the name 'Riisager-Mollen'.
Wind-Matic is affiliated to the HOLEC-Group.

Wind-Matic A/S,
Industrivej Nord 15,
Birk,
DK-7400 Herning,
Denmark
Telephone: (07) 127700
Telex: 62250

Wind Power Systems Storm Master Model 12

PERFORMANCE AND OPERATIONAL DATA

Rated electrical power 40 000 watt
Rated output voltage 440–480 volt a.c.
(operated in parallel with 3-phase, 60 Hz network)
Cut-in wind speed 6 m/s
Rated wind speed 18 m/s

PHYSICAL DATA

Turbine
Type: Horizontal axis, downwind turbine operating freely in yaw
Dimensions
diameter 12 m
Blades
Construction: Fibreglass composite with foam core and finished with linear polyurethane
Number: 3
Control: Centrifugally operated blade pitch control
Generator: Induction generator
Mechanical transmission: 13.9:1 ratio two-stage helical gearbox
System weight: 840 kg

PRODUCTION

This supercedes earlier Model 10 (shown)

MISCELLANEOUS

Electrical brake actuated by vibration switch or manually, and capable of stopping rotation within 10 revolutions.
Annual maintenance comprising lubrication of generator and gearbox, greasing of pitch change bearings and turntable bearings.
One year warranty on materials and workmanship.
Unarco-Rohn model SSV, 20 m high is standard tower.

Wind Power Systems Inc.,
8630 Production Avenue,
San Diego,
California 92121,
USA
Telephone: (714) 566 1806

Windpumpen-Zentral HD312

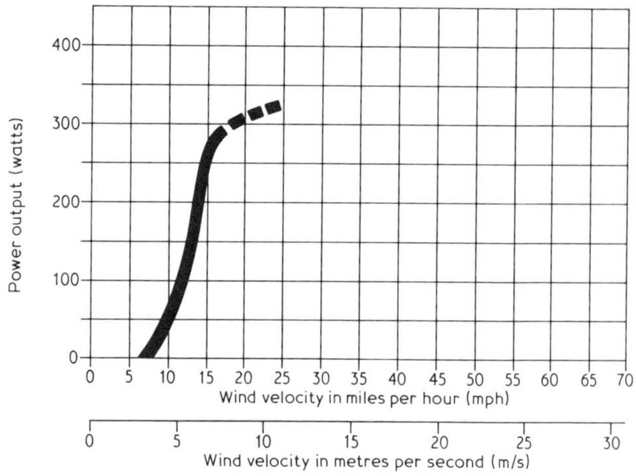

PERFORMANCE AND OPERATIONAL DATA

Rated electrical power	300 watt
Rated output voltage	12 volt d.c.
Cut-in wind speed	2.8 m/s
Rated wind speed	8.2 m/s
Shut-down wind speed	manual
Survival wind speed	30 m/s
Shaft speed at rated output	325 rev/min
Maximum shaft speed	366 rev/min

PHYSICAL DATA

Turbine

Type: Horizontal axis, upwind type with tail vane

Dimensions

diameter 2.36 m

Blades

Construction: Polyurethane hardened foam with tempered surface

Number: 2

Control: Centrifugally operated blade pitch control

Generator: a.c. generator (rectified)

Mechanical transmission: Direct coupling to turbine

System weight: 150 kg (including tower)

PRODUCTION

First production: 1972

Production quantity to date: About 100 (1980)

MISCELLANEOUS

4 m lattice tower is offered as standard.

6 monthly greasing of bearings and lubrication of braking system, annual check of brushes.

VARIANT

Model HD324 is rated 24 volt d.c. with a cut-in wind speed of 2.4 m/s, rated wind speed of 6.3 m/s and shaft speed at rated output 235 rev/min.

Windpumpen-zentrale,
Luetthoern 51,
D-2330 Eckenfoerde,
West Germany
Telephone: 04351 42024

Alfa

PERFORMANCE AND OPERATIONAL DATA

Rated electrical power	750 watt
Rated output voltage	12/24 volt d.c.
Cut-in wind speed	3.5 m/s
Rated wind speed	8.0 m/s
Shut-down wind speed	20 m/s
Survival wind speed	50 m/s
Shaft speed at rated output	260 rev/min
Maximum shaft speed	400 rev/min

PHYSICAL DATA

Turbine
Type: Horizontal axis, upwind with tail vane
Dimensions
diameter 2.6 m
Blades
Construction: Polyurethane with internal steel ribs
Number: 3
Control: Fixed pitch turbine, aerodynamic tipbrake
Generator: Synchronous a.c. generator
Mechanical transmission: Speed-increasing gearbox

PRODUCTION

First production: January 1980
Production quantity to date: 8 (September 1982)

MISCELLANEOUS

Three year guarantee

VARIANTS

The 750 watt rating listed here corresponds to type BW 1000. Variants BW 2000 (rated 1700 watt) and BW 3000 (rated 2600 watt) are also available. These have 11 m/s and 16 m/s rated wind speeds, respectively, and the BW 3000 has 4 m/s cut-in wind speed.

Windsol Ltd,
18 Chatzopoulou Street, Kallithea,
Athens,
Greece
Telephone: 92 32 943

Beta

PERFORMANCE AND OPERATIONAL DATA

Rated electrical power	5000 watt
Rated output voltage	220 volt, variable
Cut-in wind speed	3.5 m/s
Rated wind speed	10.0 m/s
Shut-down wind speed	20 m/s
Survival wind speed	50 m/s
Shaft speed at rated output	150 rev/min
Maximum shaft speed	180 rev/min

PHYSICAL DATA

Turbine
Type: Horizontal axis, upwind type with tail vane
Dimensions
diameter 6.0 m
Blades
Construction: Reinforced fibreglass
Number: 3
Control: Blade pitch control
Generator: Synchronous or induction generator
Mechanical transmission: Speed-increasing gearbox

PRODUCTION

First production: January 1981
Production quantity to date: 4 (September 1982)

MISCELLANEOUS

Three year guarantee.

Windsol Ltd.,
18 Chatzopoulou Street, Kallithea,
Athens,
Greece
Telephone: 92 32 943

Yamada Tempu 1

PERFORMANCE AND OPERATIONAL DATA

Rated electrical power	1500 watt
Rated output voltage	12/24/110 volt d.c.
Cut-in wind speed	3.5 m/s
Rated wind speed	8 m/s
Shut-down wind speed	50 m/s
Survival wind speed	100 m/s
Maximum shaft speed	250 rev/min

PHYSICAL DATA

Turbine
Type: Horizontal axis, upwind with tail vane
Dimensions
diameter 3.8 m
Blades
Construction: Wood with epoxy-polyurethane coating
Number: 2
Control: Fixed pitch turbine; thrust in high wind speed tilts turbine upwards into horizontal plane
Generator: Shunt-wound d.c. generator
Mechanical transmission: Direct coupling to turbine
System weight: 300 kg

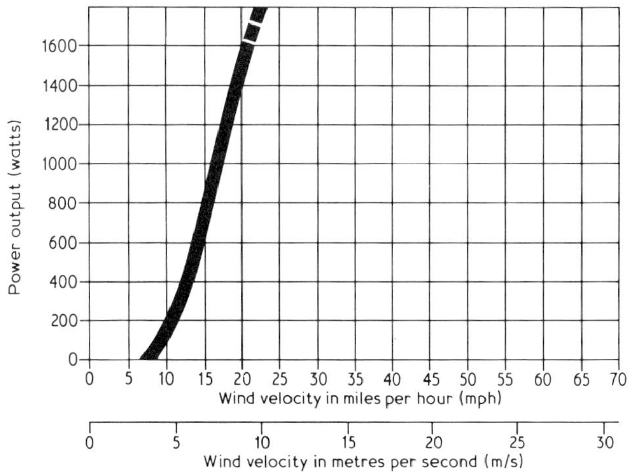

PRODUCTION

First production: 1977
Production quantity to date: About 25 (1979)

MISCELLANEOUS

11 m tower comprising three concrete poles is offered.
3–5 yearly maintenance on brushes.
Many thousand units of an obsolete type were produced in the 1950s and 1960s.
Production began with a double turbine unit.

Motohiro Yamada,
Hinomaru Pro Co. Ltd.,
1–15 Yagouchi-machi,
Edogawa-ku,
Tokyo,
Japan
Telephone: (03) 679 6623

7.1.2 *One-off small/medium wind turbine generators now built*

(including some probably commercial wind generators where available information was inadequate for inclusion in Section 7.1.1)

Manufacturer	Model	Diameter (m)	Rated power (kW)	Rated wind speed (m/s)	Type	Number of blades
Aerowatt, 37, Rue Chanzy, 75011 Paris, France	18 KFP 7 100 KFP 13	18.0 18.0	18.0 kVA 100.0 kVA	7.0 13.0	Horizontal axis upwind, with tail vane	2
Agroindustrias SRL, Pueyrredon 1534- Piso 4 'A', 1118 Buenos Aires, Argentina	Agrocom Agrocom Hidrolux	1.82 2.30	0.42 0.75 1.25		Horizontal axis, similar 'WINCO'	2
Alcoa Allied Products, Alcoa Technical Center, Alcoa Center, PA 15069, USA	217806 453011 634214	8.2×5.5 13.7×9.2 19.2×12.8	8.0 22.0 57.0	13.8	Vertical axis troposkien type	3
Balance Engineering, 135 Cambridge Road, New Wimpole, Near Royston, Hertfordshire, UK	Ro–Tow–Vane	0.52	0.006		Horizontal axis with tail vane	15
Bouma Windenergie BV, Stevinstraat 11, 1704 RN Heerhugowaard, Netherlands	—	16.0	55	12	Horizontal axis upwind, motorized orientation	3
Böwe Maschinenfabrik GmbH, Haunstatter Str. 112, 8900 Augsburg, West Germany	W112	12.0	10	8	Horizontal axis	1
Dakota Wind & Sun Ltd, PO Box 1781, Aberdeen, South Dakota 57401, USA	SI 10	8.5	10.0	8.9	Horizontal axis, variable pitch, with tail vane	3
Dana Vindkraft, Mejerivej 3, 8763 Rask Mølle, Denmark	HD II	11.1	22.0	10.5	Horizontal axis, variable pitch with fan-tails	3
Danregn Vindkraft A/S, Fabriksvej 7, 7330 Brande, Denmark	Bonus 22/3 Bonus 30/5.5 Bonus 55/15	10.2 10.2 15.3	22 30 55	— — 12.5	Horizontal axis, upwind, fixed pitch blades	3
Dansk Vindkraft Industri ApS, Vendevej 6, Buresø, DK-3550, Denmark		5.5×9.0	13.0	11.0	Giromill	3
Denki Seiko		4.0×4.0	4.0	12.0	Vertical axis troposkien type	2

Manufacturer	Model	Diameter (m)	Rated power (kW)	Rated wind speed (m /s)	Type	Number of blades
Dornier-System GmbH, Postfach 1360, 7990 Friedrichshafen 1, West Germany		6.5×5.5	5.0	12.0	Vertical axis troposkien type	3
Energy Unlimited Inc, 2 Aldwyn Center, Villanova, PA 19085, USA	Windworker 10	10.0	10.0	10.0	Horizontal axis downwind type with variable pitch	3
Enertech Corporation, PO Box 420, Norwich, Vermont 05055, USA	Enertech 4000	6.0	4.0	10.7	Horizontal axis downwind type with fixed pitch	3
Ingeniørfirma Ditlev v. Eriksen, Folehavevej 59, 2970 Hørsholm, Denmark	IDE Windmill	11.28	15.0		Horizontal axis downwind type with variable pitch	3
Fokker-VFW, Schiphol-Oost, PO Box 7600, Netherlands	VAWT 5.3	5.3	15.0	16.2	Vertical axis troposkien type	2
Grumman Energy Systems Inc, 4175 Veterans Memorial Highway, Ronkonkoma, New York 11779, USA	Windstream 33	10.15	15.0	10.7	Horizontal axis downwind type with variable pitch	3
Jayward Systems Ltd, 15 Westfield Grove, Finchfield, Wolverhampton WV3 8EX, UK	Windcharge 5	0.41	0.005	12	Horizontal axis, upwind type with tail vane. Fixed pitch	7
	Windcharge 60	0.69	0.06	18		
Lagerwey v.d. Loenhorst BV, Garderbroekerweg 175, 3774 ID Kootwijkerbrock, Netherlands	—	10.6	11	13	Horizontal axis, upwind, variable pitch, with fan tail	3
Sir Henry Lawson-Tancred & Sons Ltd, Aldborough Manor, Boroughbridge, N. Yorkshire YO5 9EP, UK		17	100		Horizontal axis, upwind type with fan-tail. Fixed pitch	3
Maschinenfabrik Augsburg-Nürnberg AG, Postfach 500620, 8000 München 50, West Germany	Aeroman 11/11	11.0	11, 14 or 22 kVA	8.5	Horizontal axis type with variable pitch and fan-tail	2
Mehrkam Energy Development Co, RD2 Box 179E, Hamburg, PA 19526, USA	4-100		100.0	13.0	Horizontal axis downwind type	4
	4-225		225.0			
Nordtank, T. Rørbaek Jensen A/S, Nyballevej 8, DK-8444 Balle, Denmark	Nordtank 90 kW	20.0	90.0	14.0	Horizontal axis, upwind type with fixed pitch	3

Manufacturer	Model	Diameter (m)	Rated power (kW)	Rated wind speed (m/s)	Type	Number of blades
Nordvestsjaellands Energivaerksted, Ringstedvej 189, 4300 Holbaek, Denmark	VM 15	15.0	55.0	12	Horizontal axis type with fixed pitch	3
Polenko BV, Remmerden 9, 3911 TZ Rhenen, Netherlands	WPS 8 WPS 11 WPS 16	8.5 11.5 16.0	15 40 60	12.0	Horizontal axis, upwind, with fixed pitch blades and fan tails	3
Praktisk Teknik AB, Box 40417, 931 04 Skelleftea, Sweden	PT-14	14 m	25	12.0	Horizontal axis, downwind with variable pitch	2
Product Development Institute Inc, 4445 Talmadge Road, Toledo, Ohio 43623, USA	Wind Jennie Model 6500	6.1	6.5	11.6	Horizontal axis, upwind type with tail vane. Fixed pitch	3
Rutland Electric Fencing Co. Ltd, Pillings Road Industrial Estate, Oakham, Rutland, Leicestershire, UK	WG 910	0.91 m	0.05	9.8	Horizontal axis, upwind type with fixed pitch blades and tail vane	6
Svensk Vindkraftindustri AB, Vettershaga 2784, 760 10 Bergshamra, Sweden	WECS 065	2.0	0.70	14	Horizontal axis, fixed pitch turbines, 'up-type'	3
Vedana Vindmöller ApS, Dag Hammarskjøldс Alle 23, 2100 København Ø, Denmark		9.0 17.0	7.5 or 15.0 30.0 or 45.0	11	Horizontal axis, upwind type with tail vane. Fixed pitch	3
Vendelbo Trapper, Gøstrup Hede 93, 9480 Løkken, Denmark	Type 15	7.0	15 kVA		Horizontal axis, upwind type with tail vane. Variable pitch	5
Volker Heindl, 2332 Koselfeld, West Germany	W10'	10.0	10.0	14	Horizontal axis, upwind, fixed pitch turbine	18
Vølund Fibreglass and Windtechnic A/S, Marsk Stigsvej 4, DK-8800 Viborg, Denmark		29.5	265.0	12	Horizontal axis, downwind type with variable pitch	3
Wellington Windpower, Wellington, Ontario, Canada		3.96	1.5		Horizontal axis, upwind with tail vane	3
Wenco SA, Ch 6533 Lumino, Via Cantonale, PO Box 7, Switzerland	W 15060E W 18060E W 18100E	15.0 18.0 18.0	60 60 100	11.3	Horizontal axis, upwind type with tail vane. Variable pitch	2

Manufacturer	Model	Diameter (m)	Rated power (kW)	Rated wind speed (m/s)	Type	Number of blades
Winflo, 15–85 Nantucket Boulevard, Scarborough, Ontario, Canada M1P 2N7	Windgen 2000	3.66	2.0	10.7	Horizontal axis, downwind type	3
WTG Energy Systems Inc, PO Box 87, 1 La Salle Street, Angola, NY 14006, USA	MP1-200	24.4	200.0	12.5	Horizontal axis, upwind type with fixed pitch and tip-flaps	3

7.1.3 *Small/medium wind turbine generators projected for development*

Manufacturer	Model	Diameter (m)	Rated power (kW)	Rated wind speed (m/s)	Type	Number of blades
Aermotor, Commerce & Exchange Street, Conway, Arkansas 72032, USA		7.6	5.0	9	Horizontal axis, fixed pitch. For network connection	3
Aero Power Systems Inc, 2398 4th Street, Berkeley, California 94710, USA	SL500 SL1500		0.5 5.0		Direct drive	
Aerowatt, 37 rue Chanzy, 75011 Paris, France	1000 FP9	3.2	1.0	9	Similar to rest of range, will replace 300 FP76 and 1200 FP 10G	2
Bristol Aerospace Ltd, PO Box 874, Winnipeg, Manitoba R3C 2S4, Canada	Licusν D	$21.0^{3}14.0$	44.0	25	Vertical axis troposkien type	2
(Licus®: registered trademark of Bristol Aerospace Ltd)						
Herman Brümmer Windkraftanlagen KG, Mühlenstrasse 1-8, 3522 Bad Karlshafen 2, West Germany	BW100 BW200	10.0 20.0	7.5 30.0	8	Horizontal axis, downwind type	3
Elektro GmbH, St Gallerstrasse 27, Winterthur, CH-8400, Switzerland			15.0		Induction generator for network connection	
Jay Carter Enterprises Inc, PO Box 684, Burkburnett, Texas 76354, USA		19.5 m	125.0		Horizontal axis, downwind type with fixed pitch	2
Naestved Maskin Elektro, Fensmarkvej 7, 4700 Naestved, Denmark	GE3 GE40 Type 80 Type 500 Type 5000		0.003 0.04 0.08 0.5 5.0	10 10 10	Savonius Horizontal axis Horizontal axis fixed pitch with tail vane	3 8 2
Polymarin B.V., Nijverheidsweg 7, 1671 GC Medemblik, Netherlands		15.0	159.0	17.5	Vertical axis troposkien type	2
Ralph Howe Marketing Ltd, Nuffield Industrial Estate, Poole, Dorset UK		3.0	0.6–0.75		Horizontal axis, upwind type with tail vane. Variable pitch	4
Windfos, Nordvestvej 4, 9000 Aalborg, Denmark		15.0 16.8	45.0 55.0		Horizontal axis, upwind type with twin fan-tails. Fixed pitch	3

7.1.4 *Large wind turbine generators built or in process of construction*

Large wind turbine generators built before 1970

Year	Country	Site	Constructor	Operator	Rated power (kW)	Rated wind speed (m/s)	Turbine diameter (m)	Number of blades	Turbine speed (rev/min)	Generator	General
1931	USSR	Yalta	Ts VEI, Central Institute of Wind Power Engineering		100	10.5	30.0	3	30	Induction	Variable pitch with auxiliary flaps. Complete tower orientates with trailing strut running on circular track
1941	USA	Grandpa's Knob, Vermont	S. Morgan Smith Co.		1250	13.5	53.3	2	26	Synchronous	Variable pitch. Coned stainless steel blades of uniform chord
1954	UK	Costa Hill, Orkney	John Brown & Co, Ltd	UK National Wind Power Committee	100	15.25	15.3	3	130	Induction	Variable pitch with individual blade articulation for coning and drag
1955	UK	St Albans then Grand Vent, Algeria	Enfield Cables Ltd	UK National WP Committee, then EGA	100	13.5	24.4	2	≤95.4	Synchronous	Variable pitch. Power transmission on the depression or Andreau principle with air turbine in the tower
1957	Denmark	Gedser, Falster	SEAS/DEF Wind Power Committee		200	15.0	24.0	3	30	Induction	Fixed pitch, regulation by stall. Overspeed protection by rotating tips. Roller chain generator drive
1957	West Germany	Stötten	Allgaier	Deutsche Studiengesellschaft Windkraft eV	100	8.0	34.0	2	42		Variable pitch. Lightweight composite blades
1958	France	St Rémy-des-Landes (Manche)	Ets. Neyrpic	EdF	132	12.5	21.2	3	56	Induction	Variable pitch, turbine downwind of tower
1958	France	Nogent-le Roi (Eure et Loire)	BEST-Romani	EdF	800	16.7	30.2	3	47.3	Synchronous	Fixed pitch, with mechanical and electrical braking. Turbine downwind of tower
1959	UK	Isle of Man	R. Smith (Horley) Ltd	ERA / Ministry of Power	100	18.5	15.2	3	75	Induction	Fixed pitch with fantail for orientation. Uniform chord extruded aluminium blades
1963	France	St Rémy-des-Landes (Manche)	Ets. Neyrpic	EdF	1000	17.0	35.0	3		Induction	Pitch varied above 650 kW. Turbine downwind of tower

Large wind turbine generators built since 1970, or presently at design or construction stage

Year of first operation	Country	Site	Machine title	Constructor	Rated power (kW)	Turbine diameter (m)	Rated wind speed (m/s)	Number of blades	General comments	Status of uncompleted machines
1975	USA	Plum Brook, Ohio	MOD O		100	38	6.5	2	Horizontal axis, initially variable pitch, downwind	Used as test bed
1977	Sweden	Kalkugnen	Experimental	SAAB–SCANIA	60	17		2	Horizontal axis downwind machine. Concrete tower	Used as test bed
1977	USA	Clayton, New Mexico	MOD OA	Westinghouse	200	38	7.7	2	Uprated version of original MOD O unit	
1977	Canada	Magdalen Islands, Quebec		DAF-Indal	224	24		2	Vertical axis troposkien machine. Aluminium blades	
1978	USA	Culebra Is., Puerto Rico	MOD OA	Westinghouse	200	38	7.7	2	Similar to previous MOD OA	
1978	Denmark	Tvind, Jutland			2000	54	14.5	3	Horizontal axis machine, concrete tower	
1979	USA	Rhode Island	MOD OA	Westinghouse	200	38	7.7	2	Similar to previous MOD OA	
1979	USA	Boone, N. Carolina	MOD 1	General Electric	2000	61	11.5	2	Horizontal axis, variable pitch, downwind	
1979	Denmark	Nibe, Jutland	NIBE A	DEFU	630	40	13	3	Horizontal axis, upwind, concrete tower	
1980	Denmark	Nibe, Jutland	NIBE B	DEFU	630	40	13	3	Horizontal axis, upwind, concrete tower	
1980	USA	Hawaii	MOD OA	Westinghouse	200	38	7.7	2	Similar to previous MOD OA, wooden blades	
1980/81	USA	Goldendale, Washington	MOD 2 (×3)	Boeing Eng. and Construction	2500	91	8.9	2	Horizontal axis, upwind with tip control. Three machines on one site	
1980	USA	California		Bendix/ Schacle	3000	50	17.9	3	Horizontal axis, upwind, orientation by rotating tower	
1980	USA	California	1238229	ALCOA	500	25	15.7	3	Vertical axis, troposkien type	
1980	USA	Oregon	1238229	ALCOA	500	25	15.7	3	As previous	

Year of first operation	Country	Site	Machine title	Constructor	Rated power (kW)	Turbine diameter (m)	Rated wind speed (m/s)	Number of blades	General comments	Status of uncompleted machines
1981	Netherlands	Petten		Fokker-VFW	300	25	13	2	Horizontal axis, variable pitch, variable speed	
1982	West Germany	Stuttgart		Voith	317	52	8.7	2	Horizontal axis, variable pitch, downwind	
	West Germany	Hamburg	GrowianI	M.A.N.	3000	100	11.8	2	Horizontal axis, variable pitch downwind	In construction
1982	Sweden	Maglarp	WTS-3	Karlskron-avarvet and Hamilton Standard	3000	78	14.2	2	Horizontal axis, variable pitch, downwind, free in yaw	
1982	Sweden	Nasudden, Gotland		KaMeWa and ERNO	2200	75	12.5	2	Horizontal axis, variable pitch, downwind	
	UK	Orkney		Wind Energy Group	250	20		2	Horizontal axis, upwind, tip control	In construction
	UK	Camarthen Bay		McAlpine and NEI	130	25	11	2	Vertical axis, variable geometry 'arrowhead' layout	In design
1982	West Germany	Bremerhaven		MBB	370	48.3	10	1	Horizontal axis, downwind, variable pitch	
	US	California		DAF-Indal	500	24	20	2	Vertical axis troposkien, aluminium blades (uprated version of Magdalen Islands unit)	In construction
	Japan	Miyakejima Island		IHI Co.	100	29.4	10	2	Horizontal axis, variable pitch	In construction
	Japan	Okineerabu Island		Mitsubishi Heavy Industries	300	33	14	3	Horizontal axis, downwind type	In construction
	US		MOD 5B	Boeing Eng. and Construction	7200	128	13.8	2	Horizontal axis, upwind, tip control	In design

7.2.1 *Commercial wind pumps*

Aermotor Division of Valley Industries Inc,
Box 1364,
Industrial Park,
Arkansas 72032,
USA
Telephone: (501) 329 9811

Also manufactured by:
Fabrica de Implementos Agricolas SAIC,
Hortiguera 1882,
Buenos Aires,
Argentina

W.D. Moore & Co,
3 Keegan Street,
O'Connor,
Western Australia 6163,
Australia

P. Andrag & Sons (Pty) Ltd,
10 Modderdam Road,
PO Box 364,
Belleville,
Cape Province,
South Africa

Model:	702X				702A				702B				702D				702E				702F			
Turbine diameter (m):	1.83				2.44				3.05				3.66				4.27				4.89			
Tower height (m):	6.4–14.35												8.24–14.35								10.08–14.35			
Pump stroke (mm):	76.2		139.7		152.4		203.2		190.5		254.0		228.6		304.8		254.0		355.6		304.8		406.4	
Cylinder diameter (mm)	Lift (m)	Capacity (m^3/h)	Lift (m)	Capacity (m^3/h)	Lift (m)	Capacity (m^3/h)	Lift (m)	Capacity (m^3/h)	Lift (m)	Capacity (m^3/h)	Lift (m)	Capacity (m^3/h)	Lift (m)	Capacity (m^3/h)	Lift (m)	Capacity (m^3/h)	Lift (m)	Capacity (m^3/h)	Lift (m)	Capacity (m^3/h)	Lift (m)	Capacity (m^3/h)	Lift (m)	Capacity (m^3/h)
44.45	52.6	0.30	39.6	0.40	75.0	0.43	56.4	0.57	113.4	0.43	85.3	0.57	170.2	0.43	128.0	0.57	243.2	0.43	182.9	0.57	405.4	0.43	304.8	0.57
47.63	48.7	0.35	36.6	0.47	70.9	0.51	53.3	0.68	105.5	0.51	79.3	0.68	158.1	0.51	118.9	0.68	227.0	0.51	170.7	0.68	372.9	0.51	280.4	0.68
50.80	38.5	0.37	29.0	0.49	56.8	0.54	42.7	0.72	87.1	0.54	65.5	0.72	129.7	0.54	97.5	0.72	186.4	0.54	140.2	0.72	304.0	0.54	228.6	0.72
57.15	31.3	0.51	23.5	0.68	45.4	0.74	34.1	0.98	68.9	0.74	51.8	0.98	101.3	0.74	76.2	0.98	145.9	0.74	109.7	0.98	239.1	0.74	179.8	0.98
63.50	26.3	0.64	19.8	0.85	38.0	0.92	28.6	1.23	56.8	0.92	42.7	1.23	85.1	0.92	64.0	1.23	121.6	0.92	91.4	1.23	198.7	0.92	149.4	1.23
69.85	22.7	0.75	17.1	1.00	32.5	1.10	24.4	1.46	48.7	1.10	36.6	1.46	73.0	1.10	54.9	1.46	105.3	1.10	79.2	1.46	172.2	1.10	129.5	1.46
76.20	19.0	0.91	14.3	1.21	27.5	1.34	20.7	1.78	40.6	1.34	30.5	1.78	62.8	1.34	47.2	1.78	89.2	1.34	67.1	1.78	145.9	1.34	109.7	1.78
82.55	—	—	—	—	—	—	—	—	35.6	1.56	26.8	2.08	52.7	1.56	39.6	2.08	75.0	1.56	56.4	2.08	123.7	1.56	93.0	2.08
88.90	14.2	1.25	10.7	1.67	20.2	1.82	15.2	2.42	30.9	1.82	23.2	2.42	46.7	1.82	35.1	2.42	64.9	1.82	48.8	2.42	107.5	1.82	80.8	2.42
95.25	—	—	—	—	—	—	—	—	26.3	2.07	19.8	2.76	39.8	2.07	29.9	2.76	58.0	2.07	43.6	2.76	93.2	2.07	70.1	2.76
101.60	10.9	1.62	8.2	2.16	15.8	2.36	11.9	3.14	23.5	2.36	17.7	3.14	34.8	2.36	26.2	3.14	50.7	2.36	38.1	3.14	81.1	2.36	61.0	3.14
107.95	—	—	—	—	—	—	—	—	20.6	2.67	15.5	3.56	30.9	2.67	23.2	3.56	44.6	2.67	33.5	3.56	73.0	2.67	54.9	3.56
114.30	8.5	2.06	6.4	2.74	12.1	2.98	9.1	3.97	18.6	2.98	14.0	3.97	27.5	2.98	20.7	3.97	39.8	2.98	29.9	3.97	64.9	2.98	48.8	3.97
120.95	—	—	—	—	—	—	—	—	—	—	—	—	24.7	3.32	18.6	4.43	35.6	3.32	26.8	4.43	56.8	3.32	42.7	4.43
127.00	6.9	2.56	5.2	3.41	10.1	3.69	7.6	4.92	15.0	3.69	11.3	4.92	22.3	3.69	16.8	4.92	32.5	3.69	24.4	4.92	52.7	3.69	39.6	4.92
146.05	—	—	—	—	—	—	—	—	—	—	—	—	16.2	4.82	12.2	6.43	24.3	4.82	18.3	6.43	40.6	4.82	30.5	6.43
152.40	—	—	—	—	6.9	5.33	5.2	7.10	10.1	5.33	7.6	7.10	15.4	5.33	11.6	7.10	22.3	5.33	16.8	7.10	34.4	5.33	25.9	7.10
177.80	—	—	—	—	—	—	—	—	7.7	7.24	5.8	9.65	11.3	7.24	8.5	9.65	16.6	7.24	12.5	9.65	26.3	7.24	19.8	9.65
203.20	—	—	—	—	—	—	—	—	5.7	9.37	4.3	12.49	8.9	9.37	6.7	12.49	12.6	9.37	9.5	12.49	20.2	9.37	15.2	12.49

NB All pumping capacities are approximate, and are based upon a 6.7–8.9 m/s wind speed.

Bobs Harries Engineering Ltd
Karamaini Estate
PO Box 40
Thika
Kenya
Telephone: Thika 47234 and 47250
Telex: 22776

Model:	'KIJITO'							
Turbine diameter (m):	3.66		4.88		6.10		7.32	
Number of blades:							24	
Tower height (m):	9.1		9.1		9.1		9.1	
Pump stroke (mm):								
Cylinder diameter (mm)	Lift (m)	Capacity (m^3/h)	Lift (m)	Capacity (m^3/h)	Lift (m)	Capacity (m^3/h)	Lift (m)	Capacity (m^3/h)
	15.3	1.42						
	30.5	0.77						
NA	45.8	0.51						
	61.0	0.38						
	91.6	0.26						
			15.3	4.37				
			30.5	2.18				
			45.8	1.46				
NA			61.0	1.09				
			91.6	0.73				
			122.1	0.55				
			152.6	0.44				
					15.3	8.59		
					30.5	4.29		
					45.8	2.86		
					61.0	2.15		
NA					91.6	1.43		
					122.1	1.07		
					152.6	0.86		
					183.2	0.72		
					213.7	0.61		
							15.3	14.84
							30.5	7.42
							45.8	4.95
							61.0	3.71
NA							91.6	2.47
							122.1	1.85
							152.6	1.48
							183.2	1.24
							213.7	1.06
							244.2	0.93

NB All pumping capacities are based upon a 3.3 m/s wind speed

Fig. 7.1 Bobs Harries 7.32 m 'Kijito' machine.

Bosman Waterbeheersing en Milieuvebetering BV
Steegjesdijk 4
Postbus 3701
3265 ZG Piershil (Z-H)
Netherlands
Telephone: (01869) 1316
Telex: 24475

Turbine diameter (m):	3.0				3.5				4.5				5.5			
Tower height (m):	6 m or 9 m				9 m or 12 m				9 m, 12 m or 15 m				12 m or 15 m			
Pump stroke (mm):	75				100				125				125			
Cylinder diameter (mm)	Lift (m)	Capacity (m^3/h) Wind speed			Lift (m)	Capacity (m^3/h) Wind speed			Lift (m)	Capacity (m^3/h) Wind speed			Lift (m)	Capacity (m^3/h) Wind speed		
		2.8 m/s	3.3 m/s	4.4 m/s		2.8 m/s	3.3 m/s	4.4 m/s		2.8 m/s	3.3 m/s	4.4 m/s		2.8 m/s	3.3 m/s	4.4 m/s
50	55	—	0.19	0.88	85	—	0.23	0.95	—	—	—	—	152	—	—	0.30
65	37	—	0.31	1.32	60	0.29	0.39	1.77	90	—	0.35	1.59	122	—	0.30	1.49
75	27	0.30	0.47	1.92	45	0.35	0.61	2.61	76	0.24	0.48	2.24	91	0.49	1.23	2.33
90	20	0.46	0.61	2.75	33	0.47	0.80	3.52	58	0.33	0.61	3.18	76	1.07	1.81	3.18
100	15	0.64	0.77	3.52	27	0.67	1.02	4.17	46	0.51	0.83	4.20	60	2.09	3.26	4.40
115	—	—	—	—	21	0.76	1.23	5.19	37	0.61	0.97	5.17	45	2.89	3.96	5.86
130	—	—	—	—	18	0.94	1.57	6.46	30	0.73	1.14	6.21	30	3.77	5.28	7.74

Bosman Waterbeheersing en Milieuvebetering BV
Steegjesdijk 4
Postbus 3701
3265 ZG Piershil (Z–H)
Netherlands
Telephone: (01869) 1316
Telex: 24475

Turbine diameter (m): 1.3 m

Number of blades: 4

Tower height (m): 4 m or 7 m

Pump stroke (mm): Centrifugal

Lift (m)	Capacity (m^3/h)					
	Wind speed					
	4 m/s	4.5 m/s	5 m/s	7 m/s	8 m/s	10 m/s
0.4	49.8	66.0	93.9	112.5	121.7	126.1
0.6	36.0	54.0	84.6	106.2	117.6	125.1
0.8	20.0	40.8	74.5	98.6	113.1	123.6
1.0	—	25.2	63.0	90.0	107.4	121.5
1.2	—	—	49.8	80.9	101.4	118.8
1.4	—	—	35.1	70.2	94.5	115.8
1.6	—	—	17.1	58.2	87.0	111.9
1.8	—	—	—	45.0	77.7	107.7
2.0	—	—	—	28.5	66.3	102.0

Bowjon
2829 Burton Avenue
Burbank
California 91504
USA
Telephone: (213) 846 2620

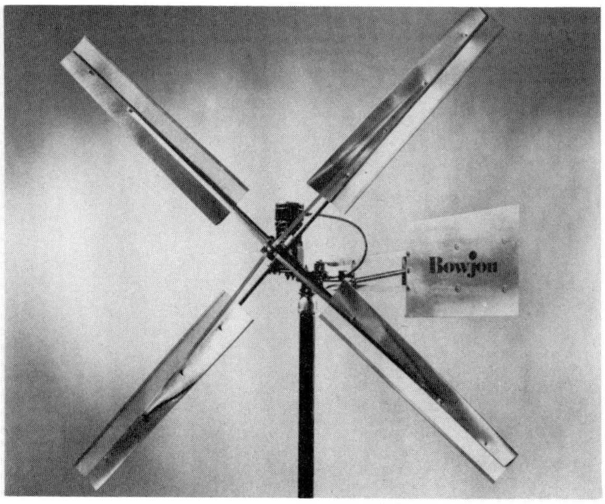

Fig. 7.2 Bowjon windmill/airlift pump.

Model: Windmill/airlift water pump

Turbine Diameter (m): 2.37

Number of blades: 4

Water line diameter (mm)	Lift (m)	Submergence (m)	Capacity (m^3/h)
12.7	30.5	22.9	0.23
12.7	30.5	30.5	0.28
12.7	61.1	61.1	0.19
19.1	15.3	15.3	0.37
19.1	15.3	22.9	0.46
19.1	22.9	22.9	0.31
19.1	22.9	30.5	0.34
19.1	30.5	15.3	0.22
19.1	30.5	22.9	0.23
25.4	9.2	15.3	0.79
25.4	9.2	18.3	0.91

NB (1) All pumping capacities are based on a 6.7 m/s wind speed.
(2) Submergence is the minimum depth of the pump below water level for satisfactory operation.

Climax Windmills (Pty) Ltd
Box 20244
General Hertzog Road
Peacehaven 1934
Vereeniging
Transvaal
South Africa
Telephone: (016) 4-5133
Telex: 8-0480

Model:	No. 6			No. 8			No. 10			No. 12			No. 14			No. 18		
Turbine diameter (m):	1.83			2.44			3.05			3.66			4.27			5.50		
Tower height (m):	6.10 or 9.16															12.21 or 15.27		
Pump stroke (mm):	127			140			178			235			235			152		
Cylinder diameter (mm)	Lift (m)	Capacity (m^3/h) Wind speed 3.25 m/s	4.44 m/s	Lift (m)	Capacity (m^3/h) Wind speed 3.25 m/s	4.44 m/s	Lift (m)	Capacity (m^3/h) Wind speed 3.25 m/s	4.44 m/s	Lift (m)	Capacity (m^3/h) Wind speed 3.25 m/s	4.44 m/s	Lift (m)	Capacity (m^3/h) Wind speed 3.25 m/s	4.44 m/s	Lift (m)	Capacity (m^3/h) Wind speed	
40	36	0.026	0.091	53	0.036	0.136	100	0.023	0.111	—	—	—	—	—	—			
45	30	0.036	0.125	40	0.049	0.185	70	0.029	0.150	104	0.044	0.189	—	—	—			
51	26	0.049	0.162	30	0.064	0.227	58	0.042	0.189	86	0.057	0.246	122	0.053	0.302			
65	15	0.075	0.246	21	0.100	0.379	39	0.066	0.302	61	0.089	0.378	100	0.066	0.360			
76	12	0.107	0.358	15	0.146	0.548	27	0.098	0.434	45	0.127	0.567	76	0.095	0.530			
90	9	0.147	0.492	11	0.189	0.737	21	0.130	0.605	35	0.174	0.776	58	0.131	0.719			
102	7	0.189	0.642	9	0.246	0.962	17	0.172	0.794	27	0.227	1.00	46	0.170	0.946			
115	6	0.246	0.812	7	0.321	1.23	12	0.208	1.00	23	0.283	1.27	38	0.208	1.21			
125	4	0.302	1.02	6	0.398	1.51	10	0.265	1.22	18	0.340	1.57	30	0.265	1.50			

Fig. 7.3 Climax Model No. 10 wind pump.

Fig. 7.4 Climax Model No. 18 wind pump.

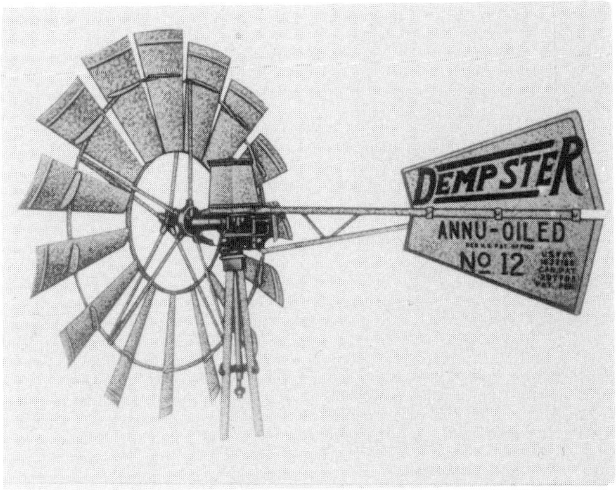

Fig. 7.5 Dempster 'Annu-oiled' No. 12 model wind pump.

Fig. 7.6 Dempster 'Annu-oiled' No. 12, 1.83 m diameter on 10.1 m tower.

Dempster Industries Inc.
PO Box 848
Beatrice
Nebraska 68310
USA
Telephone: (402) 223 4026

Model: 'Annu-oiled' windmill										
	No. 12		No. 12A		No. 12		No. 12		No. 12	
Turbine diameter (m):	1.83		2.44		3.05		3.66		4.27	
Number of blades:	15		15		24		18		18	
Tower height (m):	6.7 or 8.5 or 10.1 or 11.9						9.2 or 12.2			
Pump stroke (mm):	127.0		190.5		190.5		304.8		304.8	
Cylinder diameter (mm)	Lift (m)	Capacity (m^3/h)	Lift (m)	Capacity (m^3/h)	Lift (m)	Capacity (m^3/h)	Lift (m)	Capacity (m^3/h)	Lift (m)	Capacity (m^3/h)
47.6	36.6	0.52	52.5	0.79	78.2	0.64	118.4	0.82	177.1	0.72
50.8	29.0	0.59	41.2	0.89	64.1	0.72	92.8	0.94	138.9	0.80
57.2	22.9	0.75	32.7	1.13	50.4	0.92	73.3	1.18	109.9	1.01
63.5	18.9	0.94	27.2	1.38	41.8	1.13	61.1	1.47	91.6	1.26
69.9	16.5	1.13	23.5	1.69	36.3	1.37	52.8	1.78	79.4	1.52
76.2	13.7	1.34	19.8	2.00	31.1	1.63	44.9	2.11	67.2	1.80
82.6	11.9	1.58	16.8	2.57	26.3	1.90	38.2	2.48	57.1	2.12
88.9	10.4	1.82	14.7	2.73	22.9	2.22	33.0	2.87	49.5	2.46
95.3	8.9	2.08	12.8	3.13	19.8	2.54	28.7	3.30	43.4	2.82
101.6	7.9	2.38	11.3	3.55	17.4	2.89	25.3	3.74	37.9	3.22

NB Pumping capacities are based upon a 6.7 m/s wind speed (1.83 m, 2.44 m and 3.05 m machines) or a 8.9 m/s wind speed (3.66 m and 4.27 m).

Guanaco International SRL
Maipu 325
1006 Buenos Aires
Argentina
Telephone: 45-9686

Turbine diameter (m):	2.44		3.05		3.66		4.89	
Cylinder diameter (mm)	Lift (m)	Capacity (m^3/h)	Lift (m)	Capacity (m^3/h)	Lift (m)	Capacity (m^3/h)	Lift (m)	Capacity (m^3/h)
50.8	—	—	70	0.80	90	0.70	160	0.80
57.2	—	—	—	—	—	—	140	0.90
63.5	40	1.00	60	1.20	70	1.20	—	—
69.9	—	—	—	—	60	1.40	100	1.20
76.2	20	1.50	40	1.80	50	1.70	80	1.50
88.9	15	2.00	30	2.40	40	2.30	—	—
95.3	—	—	—	—	—	—	70	1.80
101.6	10	2.60	20	3.00	30	2.95	50	2.90
114.3	—	—	15	3.80	20	4.00	40	3.70
127.0	5	4.50	10	4.70	15	4.60	30	4.50
152.4	—	—	—	—	10	6.40	20	6.00
177.8	—	—	5	9.10	5	11.50	15	8.50
254.0	—	—	—	—	—	—	10	12.80
304.8	—	—	—	—	—	—	5	25.00

NB All pumping capacities are based upon a 5.6 m/s wind speed.

The Heller-Aller Co.
Perry and Oakwood Streets,
Napoleon
Ohio 43545
USA

Model:	Baker 6		Baker 8		Baker 10		Baker 12	
Turbine diameter (m):	1.83		2.44		3.05		3.66	
Pump stroke (mm):	107.9		134.9		161.9		200.0	
Cylinder diameter (mm)	Lift (m)	Capacity (m^3/h)	Lift (m)	Capacity (m^3/h)	Lift (m)	Capacity (m^3/h)	Lift (m)	Capacity (m^3/h)
41.3	38.2	0.55	—	—	—	—	—	—
44.5	—	—	45.8	1.00	76.3	0.98	91.6	0.91
47.6	—	—	38.2	1.09	61.1	1.18	—	—
50.8	30.5	0.68	30.5	1.14	45.8	1.27	76.3	1.48
57.2	15.3	0.91	22.9	1.59	—	—	—	—
63.5	10.7	1.09	15.3	2.05	30.5	2.10	45.8	2.39
76.2	7.6	1.59	10.7	3.28	15.3	3.19	30.5	2.73
88.9	—	—	7.6	4.10	10.7	4.21	—	—
101.6	—	—	—	—	7.6	5.69	22.9	5.12
114.3	—	—	—	—	—	—	15.3	6.49
127.0	—	—	—	—	—	—	10.7	7.40
152.4	—	—	—	—	—	—	7.6	10.93

NB All pumping capacities are approximate, and are based upon a 6.7 m/s wind speed.

Fig. 7.7 Heller-Aller 'Baker' wind pump.

Fritz Hüllmann-Windkraftanlagen
In De Horn 1,
D-2082 Tornesch
West Germany
Telephone: 0 41 22/52350

Model:	HM 36W						HM 70W					
Turbine diameter (m):	5.0						9.0					
Number of blades:	8						10					
Tower height (m):	9.0						12.0					
Pump stroke (mm):	50–200						200–400					
Cylinder diameter (mm)	Lift (m)	Capacity (m^3/h) Wind speed					Lift (m)	Capacity (m^3/h) Wind speed				
		4 m/s	6 m/s	8 m/s	10 m/s	12 m/s		4 m/s	6 m/s	8 m/s	10 m/s	12 m/s
100	4	8.1	15.1	24.3	34.5	35.0						
100	6	5.1	9.5	15.9	23.0	23.2						
100	8	4.0	7.3	12.1	17.5	17.5						
100	10	2.5	5.0	9.0	13.3	13.3						
180							4	23.4	33.1	47.2	68.1	70.0
180							6	16.0	23.6	33.9	47.5	49.0
180							8	12.3	17.6	25.6	35.2	35.8
180							10	9.7	13.4	20.0	29.4	30.0

Fig. 7.8 Fritz Hüllmann-Windkraftanlagen model HM 36W.

Fig. 7.9 Fritz Hüllmann-Windkraftanlagen model HM 70W.

Lubing Maschinenfabrik,
Postfach 110,
D-2847 Barnstorf
West Germany
Telephone: (05442) 625/6/7
Telex: 09-41212

Model: MO13-3 — *Turbine diameter (m):* 1.3 — *Number of blades:* 3 — *Tower height (m):* 3.0 — *Pump stroke (mm):*

Cylinder diameter (mm)	Lift (m)	Capacity (m^3/h) Wind speed 3.0 m/s	4.0 m/s	5.0 m/s	6.0 m/s	7.0 m/s	8.0 m/s
35	6.0	0.07	0.10	0.12	0.16	0.19	0.21
40							
50							
65							
90							
115							

Model: MO15-4 — *Turbine diameter (m):* 1.5 — *Number of blades:* 4 — *Tower height (m):* 3.7 — *Pump stroke (mm):*

Cylinder diameter (mm)	Lift (m)	Capacity (m^3/h) Wind speed 3.0 m/s	4.0 m/s	5.0 m/s	6.0 m/s	7.0 m/s	8.0 m/s
35	7.0	0.14	0.18	0.24	0.33	0.38	0.44
40							
50							
65							
90							
115							

Model: MO15-6 — *Turbine diameter (m):* 1.5 — *Number of blades:* 6 — *Tower height (m):* 3.7 or 6.7 — *Pump stroke (mm):*

Cylinder diameter (mm)	Lift (m)	Capacity (m^3/h) Wind speed 3.0 m/s	4.0 m/s	5.0 m/s	6.0 m/s	7.0 m/s	8.0 m/s
35	16.0	0.11	0.15	0.19	0.27	0.31	0.31
40	13.0	0.14	0.18	0.24	0.33	0.38	0.38
50	7.0	0.21	0.32	0.37	0.52	0.60	0.60
65							
90							
115							

Model: MO22-3 — *Turbine diameter (m):* 2.2 — *Number of blades:* 3 — *Tower height (m):* 7.0 or 10.1 or 13.1 — *Pump stroke (mm):* 200

Cylinder diameter (mm)	Lift (m)	Capacity (m^3/h) Wind speed 3.0 m/s	4.0 m/s	5.0 m/s	6.0 m/s	7.0 m/s	8.0 m/s
35	39.7	0.14	0.20	0.30	0.46	0.53	0.61
40	30.5	0.17	0.26	0.40	0.60	0.70	0.80
50	20.2	0.27	0.40	0.62	0.93	1.10	1.25
65	12.2	0.45	0.64	0.98	1.57	1.83	2.12
90	6.1	0.87	1.29	1.97	3.14	3.67	4.23
115	3.8	1.36	1.99	3.11	4.88	5.63	6.46

MBP (WA) Pty Ltd
Salvado Road, Wembley
Western Australia 6014
Telephone: (09) 381 3444
Telex: AA93413

Model:	Type 'M'									
Turbine diameter (m):	1.83		2.44		3.05		3.66		4.27	
Tower height (m):	6.10 or 7.62 or 9.14 or 12.20								9.14 or 12.20	
Pump stroke (mm):	101.6–127.0		127.0–158.7		152.4–177.8		152.4–177.8		184.1	
Cylinder diameter (mm)	Lift (m)	Capacity (m^3/h)	Lift (m)	Capacity (m^3/h)	Lift (m)	Capacity (m^3/h)	Lift (m)	Capacity (m^3/h)	Lift (m)	Capacity (m^3/h)
50.8	21.6	0.21	38.7	0.25	80.8	0.29	89.0	0.31	117.4	0.32
63.5	14.9	0.33	26.2	0.40	62.8	0.46	71.0	0.49	86.0	0.51
69.8	12.2	0.40	22.0	0.48	49.1	0.56	60.7	0.60	74.7	0.62
76.2	10.4	0.47	18.9	0.56	42.4	0.66	51.5	0.70	61.3	0.72
88.9	8.5	0.65	13.4	0.78	30.2	0.91	42.4	0.97	51.5	1.00
101.6	7.6	0.83	10.7	1.00	22.3	1.17	34.1	1.23	40.8	1.29

NB Pumping capacities are based upon a 2.7–3.6 m/s wind speed.

Pembrokeshire Engineering Services Ltd,
Keeston House,
Keeston,
Nr Haverfordwest,
Pembs,
Dyfed,
UK
Telephone: Camrose 348

Model:	Bean Hill 2/27		Bean Hill 3/88	
Turbine diameter (m):	1.98		3.05	
Number of blades:			18	
Tower height (m):	4.88		6.11	
Pump stroke (mm):				
Cylinder diameter (mm)	Lift (m)	Capacity (m^3/h)	Lift (m)	Capacity (m^3/h)
	18.3	1.33	18.3	3.79
	61.1	0.47	61.1	1.33

Fig. 7.10 Pembrokeshire Engineering Services 'Bean Hill' 3/88.

Reymill Steel Products
Sta. Rosa,
Nueva Ecija,
Philippines

Turbine diameter (m):	3.05, 3.66 or 4.27			
Pump stroke (mm):	101.6–152.4		152.4–203.2	
Cylinder diameter (mm)	Lift (m)	Capacity (m^3/h)	Lift (m)	Capacity (m^3/h)
44.5	—	—	15.3–91.6	0.37–0.61
76.2	—	—	15.3–91.6	1.06–1.78
101.6	3.05–18.3	1.14–3.03	—	—

Fig. 7.11 Reymill.

Toowoomba Foundry Pty Ltd
259 Ruthven Street
PO Box 109
Toowoomba
Queensland
Australia 4350
Telephone: (076) 32 3122
Telex: 40046

Fig. 7.12 Southern Cross 'IZ' pattern

Fig. 7.13 Southern Cross 'Seneschal'.

Model:	Southern Cross 'IZ'										Southern Cross 'Seneschal'											
Turbine diameter (m):	1.83		2.44		3.05		3.66		4.27		5.19				6.41				7.63			
Tower height (m):	7.6–18.3		9.2–18.3		9.2–18.3		12.2–18.3		12.2–18.3		10.7–18.3				12.2–18.3				12.2–18.3			
Pump stroke (mm):	133.4		146.1		165.1		184.2		184.2		177.8		203.2		209.6		254.0		241.3		304.8	
Cylinder diameter (mm)	Lift (m)	Capacity (m^3/h)	Lift (m)	Capacity (m^3/h)	Lift (m)	Capacity (m^3/h)	Lift (m)	Capacity (m^3/h)	Lift (m)	Capacity (m^3/h)	Lift (m)	Capacity (m^3/h)	Lift (m)	Capacity (m^3/h)	Lift (m)	Capacity (m^3/h)	Lift (m)	Capacity (m^3/h)	Lift (m)	Capacity (m^3/h)	Lift (m)	Capacity (m^3/h)
44.5	22.9	0.40	40.3	0.44	72.1	0.44	96.2	0.46	135.3	0.40	146.6	0.75	128.2	0.86	—	—	—	—	—	—	—	—
50.8	18.3	0.53	33.2	0.53	60.2	0.55	80.3	0.62	113.0	0.53	126.7	0.97	109.9	1.13	—	—	—	—	—	—	—	—
57.2	15.6	0.68	28.1	0.75	50.7	0.71	67.8	0.77	95.3	0.66	109.9	1.24	94.7	1.41	175.6	1.19	143.5	1.43	—	—	—	—
63.5	13.1	0.79	23.5	0.93	43.1	0.88	57.7	0.95	80.9	0.82	96.2	1.54	82.4	1.77	152.7	1.46	125.2	1.79	216.8	1.41	161.8	1.81
69.9	11.3	1.01	20.2	1.10	36.9	1.06	49.5	1.15	69.6	0.99	84.0	1.85	73.3	2.12	134.4	1.77	109.9	2.16	193.9	1.72	145.0	2.16
76.2	9.8	1.21	17.4	1.32	32.1	1.28	42.7	1.37	60.2	1.15	73.3	2.21	64.1	2.54	119.1	2.12	97.7	2.56	175.6	2.05	129.8	2.60
82.6	8.2	1.41	15.3	1.54	28.1	1.50	37.6	1.61	52.5	1.35	62.6	2.60	55.0	2.98	105.3	2.47	87.0	3.00	157.3	2.41	117.6	3.05
88.9	7.3	1.63	13.4	1.79	24.7	1.72	33.0	1.85	46.1	1.59	53.4	3.02	47.3	3.44	94.7	2.87	76.3	3.49	140.5	2.78	106.9	3.53
101.6	5.8	2.14	10.4	2.34	19.5	2.25	26.0	2.43	36.3	2.03	41.2	3.95	36.6	4.50	73.3	3.75	61.1	4.55	106.9	3.64	85.5	4.61
108.0	5.2	2.41	9.5	2.65	17.4	2.56	23.2	2.74	32.7	2.30	36.6	4.46	32.1	5.08	64.1	4.24	53.4	5.14	94.7	4.10	74.8	5.21
114.3	4.6	2.69	8.5	2.96	15.6	2.87	20.8	3.07	29.3	2.60	32.1	4.99	29.0	5.69	58.0	4.77	47.3	5.78	85.5	4.61	67.2	5.83
127.0	3.7	3.33	7.0	3.66	12.8	3.53	17.1	3.80	24.1	3.18	26.0	6.18	22.9	7.06	47.3	5.87	38.2	7.13	68.7	5.69	55.0	7.19
152.4	—	—	4.9	5.30	9.2	5.08	12.2	5.47	17.1	4.59	18.3	8.83	16.2	10.15	32.1	8.47	26.9	10.24	47.3	8.21	38.2	10.37
203.2	—	—	—	—	—	—	7.0	9.71	9.8	8.30	10.4	15.89	9.2	18.01	18.3	15.01	15.3	18.18	26.9	14.57	21.4	18.45
254.0	—	—	—	—	—	—	—	—	—	—	—	—	—	—	11.6	23.48	9.8	28.42	16.8	22.77	13.7	28.78
304.8	—	—	—	—	—	—	—	—	—	—	—	—	—	—	8.2	33.81	6.7	41.05	11.9	32.84	9.5	41.49
355.6	—	—	—	—	—	—	—	—	—	—	—	—	—	—	—	—	—	—	8.5	45.02	7.0	56.94

NB All pumping capacities are based on a 8.0 m/s wind speed; output is reduced by 25% in 5.4–7.1 m/s wind speed and by 50% in 3.6–4.5 m/s wind speed.

Wakes & Lamb Ltd
Millgate Works
Newark
Notts NG24 4XB
UK
Telephone: (0636) 704464

Model:	Newark		Newark		Newark		Newark		Newark		Newark	
Turbine diameter (m):	1.83		2.44		3.05		3.66		4.28		4.89	
Pump stroke (mm):	152.4		165.1		177.8		203.2		228.6		228.6	
Cylinder diameter (mm)	Lift (m)	Capacity (m^3/h)	Lift (m)	Capacity (m^3/h)	Lift (m)	Capacity (m^3/h)	Lift (m)	Capacity (m^3/h)	Lift (m)	Capacity (m^3/h)	Lift (m)	Capacity (m^3/h)
50.8	22.9	0.48	30.5	0.54	45.8	0.58	76.3	0.55	106.9	0.50	122.1	0.50
57.2	—	—	—	—	—	—	—	—	91.6	0.63	106.9	0.63
63.5	15.3	0.76	22.9	0.85	38.2	0.91	61.1	0.87	76.3	0.77	91.6	0.77
76.2	7.6	1.09	15.3	1.22	30.5	1.31	45.8	1.25	61.1	1.13	76.3	1.13
88.9	—	—	—	—	22.9	1.79	38.2	1.70	45.8	1.53	61.1	1.53
101.6	—	—	7.6	2.16	15.3	2.33	30.5	2.22	38.2	2.00	45.8	2.00
114.3	—	—	—	—	7.6	3.64	22.9	2.81	30.5	2.53	38.2	2.53
127.0	—	—	—	—	—	—	15.3	3.46	22.9	3.12	30.5	3.12
152.4	—	—	—	—	—	—	—	—	15.3	4.51	22.9	4.51
177.8	—	—	—	—	—	—	7.6	6.83	—	—	—	—
203.2	—	—	—	—	—	—	—	—	7.6	7.97	15.3	7.97
254.0	—	—	—	—	—	—	—	—	—	—	7.6	12.39

Fig. 7.14 Wakes & Lamb 'Newark' 4.88 m diameter on 9.2 m tower.

Sidney Williams & Co (Pty) Ltd
Williams Parade
PO Box 22
Dulwich Hill
New South Wales
Australia 2203
Telephone: 560 4000

Model: Comet

Turbine diameter (m):	2.44		3.05		3.66		4.27		4.89		5.50		6.10		6.71		7.32		8.23		9.14	
Pump stroke (mm):	76.2		88.9		101.6		127.0		203.2		203.2		254.0		254.0		304.8		355.6		355.6	
Cylinder diameter (mm)	Lift (m)	Capa-city (m^3/h)	Lift (m)	Capa-city (m^3/h)	Lift (m)	Capa-city (m^3/h)	Lift (m)	Capa-city (m^3/h)	Lift (m)	Capa-city (m^3/h)	Lift (m)	Capa-city (m^3/h)	Lift (m)	Capa-city (m^3/h)	Lift (m)	Capa-city (m^3/h)	Lift (m)	Capa-city (m^3/h)	Lift (m)	Capa-city (m^3/h)	Lift (m)	Capa-city (m^3/h)
44.5	28.9	0.15	48.8	0.16	67.1	0.16	94.5	0.18	—	—	—	—	—	—	—	—	—	—	—	—	—	—
50.8	22.9	0.20	39.6	0.21	53.3	0.21	76.2	0.24	97.5	0.34	125.0	0.31	—	—	—	—	—	—	—	—	—	—
57.2	18.3	0.25	32.0	0.27	42.7	0.27	60.9	0.30	80.8	0.43	107.0	0.39	146.0	0.44	—	—	—	—	—	—	—	—
63.5	15.8	0.31	25.9	0.33	36.6	0.34	51.8	0.37	67.1	0.54	91.4	0.48	126.0	0.54	152.0	0.48	198.0	0.54	—	—	—	—
69.9	13.7	0.38	21.9	0.40	30.5	0.41	42.7	0.45	57.9	0.65	76.2	0.59	110.0	0.65	131.0	0.59	174.0	0.65	218.0	0.72	300.0	0.64
76.2	11.6	0.45	18.3	0.47	25.9	0.48	36.6	0.54	48.8	0.77	64.0	0.70	93.0	0.78	116.0	0.70	149.0	0.77	189.0	0.82	256.0	0.75
82.6	9.8	0.53	15.2	0.56	21.9	0.57	32.0	0.63	41.1	0.91	54.9	0.81	79.2	0.91	99.1	0.81	131.0	0.91	158.0	0.97	198.0	0.88
88.9	8.5	0.62	13.7	0.65	19.2	0.65	27.4	0.73	36.0	1.05	47.2	0.95	68.6	1.06	85.3	0.95	113.0	1.05	143.0	1.12	174.0	1.02
101.6	6.7	0.80	10.7	0.85	14.6	0.86	21.3	0.96	28.7	1.37	37.8	1.24	54.9	1.38	67.1	1.24	85.3	1.37	110.0	1.47	139.0	1.33
108.0	5.5	0.90	9.1	0.96	13.1	0.97	18.9	1.07	25.3	1.55	33.5	1.39	48.8	1.56	61.0	1.40	76.2	1.55	99.0	1.66	130.0	1.51
114.3	4.6	1.07	8.2	1.07	11.6	1.08	16.8	1.19	22.9	1.73	28.9	1.56	42.7	1.75	51.9	1.56	67.1	1.73	86.9	1.86	113.0	1.69
127.0	3.7	1.25	6.7	1.33	9.1	1.36	13.7	1.49	18.3	2.14	22.9	1.93	33.5	2.16	41.1	1.93	53.3	2.16	70.1	2.29	89.9	2.08
152.4	—	—	4.6	1.89	6.1	1.92	9.1	2.14	12.2	3.09	15.2	2.78	21.3	3.11	29.0	2.78	36.6	3.09	47.2	3.30	64.0	2.99
177.8	—	—	3.4	2.50	4.6	2.59	6.7	2.92	9.1	4.17	11.6	3.79	17.1	4.24	22.9	4.74	28.0	4.21	30.5	4.49	42.7	4.09
203.2	—	—	—	—	3.7	3.41	5.2	3.79	7.3	5.49	9.1	4.92	13.7	5.49	18.3	6.06	21.3	5.40	24.4	5.87	33.5	5.30
254.0	—	—	—	—	—	—	3.0	5.87	4.3	8.33	5.5	7.58	7.9	8.52	10.7	9.47	12.8	8.33	15.8	8.90	21.9	8.14
304.8	—	—	—	—	—	—	—	—	—	—	3.7	10.99	5.2	12.31	7.0	13.64	9.1	12.12	11.6	13.07	15.8	11.93
381.0	—	—	—	—	—	—	—	—	—	—	—	—	3.6	18.94	4.9	20.84	5.8	18.75	7.0	20.27	9.8	18.56
457.2	—	—	—	—	—	—	—	—	—	—	—	—	—	—	—	—	4.0	27.47	5.2	29.17	7.0	26.52

NB All pumping capacities are based upon an average 2.5 m/s wind speed.

7.2.2 *Other wind pump manufacturers known to be active*

Alternative Energy Ltd	Woodford Co. Galway Ireland Telephone: Woodford 52 Telex: 5339
African Gate & Fence Works	PO Box 7554 Johannesburg 2000 South Africa
S.A. Bruno	Route du Mans Bonchamps-les-Laval 53210 Argentre France
Hermann Brümmer Windkraftanlagen KG	Mühlenstrasse 1-8 3522 Bad Karlshafen 2 West Germany Telephone: 05672/8 20
Éoliennes Humblot	8 rue d'Alger Coussey 88300 Neufchateau France Telephone: (29) 94 09 09
GERA Bombas e Implementos Ltda	Rua Brigadiero Machado 243 PO Box 10664 03050 São Paulo Brazil
Gjellerup Smed	Kirkevaenget 3 7400 Herning Denmark Telephone: 07-116073
Maquinas Agricolas Fortuna Ltda	Rua João Adolfo 118 Sala 711 Sao Paulo Brazil Telephone: 36-5160 and 239-4467 Telex: 01121724
Millville Windmills Inc.	PO Box 32 (10335 Old 44 Drive) Millville California 96062 USA Telephone: (916) 547–4302
Sjørslev Maskinforretning	Sjørslev 8620 Kjellerup Denmark

Sparco Pumps	A/S Naesbjerg 6800 Varde Denmark
Wadler Manufacturing Co. Inc.	Rt 2 Box 76 Galena Kansas 66739 USA Telephone: (316) 783 1355
Windpumpen-Zentrale	Lütthörn 51 D-2330 Eckenförde West Germany Telephone: 04351 42024

7.3 Anemometers

Manufacturer	Model	Type	Wind speed range	Maximum wind speed	Response	Accuracy	Temperature range (°C)	Speed transducer	Direction transducer	Output instrument
Aanderaa Instruments, Fanaveien 13, PO Box 160, 5051 Nesttun, Norway Telephone: Bergen 2740 30	2593	Three cup rotor	—	0·3–60 m/s	—	± 2%	− 40 to + 50	—	—	Form part of an automatic weather station, with magnetic tape recording. Tape reader 2650 and ABC80 displayer (CRT, keyboard and cassette program loader) and telemetry are available
	2053	Direction vane	—	—	—	—	—	—	—	
Approach Fish Inc, RT 1, Box 620B, Ringgold, Virginia 24586, USA Telephone: 804 793 2828	150	Handheld kite	3–20 m/s 15–63 mph	—	—	—	—	Spring tension	—	Spring gauge in launching rod. Standard kite suitable for altitudes 2–250 m, high altitude version 2–400 m. Motorized reel version (150A) available.
	151	Tethered kite	3–25 m/s	—	—	—	—	Strain gauge	—	Recorded on Rustrac recorder. Motorized reel version (151A) available.
	152	Three axis recording kite	3–25 m/s	—	—	2% on speed 2% on altitude 1% on direction (in absence of shear)	—	Strain gauge	—	Two channel analogue recorder, switchable velocity/direction; altitude/zero. 0–10 volt d.c. output each channel. Microprocessor based data acquisition system (153) available
Brookes and Gatehouse Ltd, Bath Road, Lymington, Hampshire SO4 9YP, UK Telephone: (0590) 74252	152	Three cup rotor	0–100 mph	—	1 s	± 4%	−10 to +60	Tacho-generator	—	Indicating instrument and or alarm facility.
	160	Three cup rotor and direction vane	0–40/80 knot	—	1 s	±5% or 1.5 knot whichever is greater (speed); ± 3° (direction)	−10 to +60	Reed-switches (twice on-off per revolution)	3 phased d.c. synchro 355° potentiometer	Indicating instruments or alarm facility, or analogue (particularly roll chart) recorders
C. F. Casella & Co Ltd, Regent House, Britannia Walk, London N1 7ND, UK Telephone: 01 253 8581 Telex: 261641	W1000	Direction vane	—	—	—	—	—	—	None	(UK Met. Office pattern IIB)
	W1010/1	Direction vane	—	—	—	—	—	—	None	(Simple inexpensive wind vane)
	W1200/2 W1204/2 W1208/2	Three cup rotor	5–80 knot	—	—	Better than 1 knot over 5–80 knot range	—	Geared odometer	—	Inclined window, nautical miles run of wind counter Inclined window, statute miles run of wind counter Inclined window, kilometres run of wind counter
	W1220 W1222 W1224	Three cup rotor	5–80 knot	—	—	Better than 1 knot over 5–80 knot range	—	Geared odometer	—	Vertical window, nautical miles run of wind counter Vertical window, statute miles run of wind counter Vertical window, kilometres run of wind counter

Manufacturer	Model	Type	Wind speed range	Maximum wind speed	Response	Accuracy	Temperature range (°C)	Speed transducer	Direction transducer	Output instrument
	W1250/2 W1252/2 W1254/2	Three cup rotor	5–80 knot	—	—	Better than 1 knot over 5–80 knot range	—	Contact each 0·1 nautical mile. Contact each 0·1 statute mile. Contact each 0·1 kilometre	—	W1258 run of wind indicator, or ten-minute mean wind speed indicators, W1260 (manual), W1262 (automatic, 200–250-volt, 50 Hz), W1264 (automatic, 110 volt, 60 Hz), or W1261/1 battery-operated audible indicator
	W1300 W1302 W1304	Three cup rotor	5–90 knot 6–100 mph 9–160 km/h	—	—	From ± ½ knot at 5 knot to ± 2 knot at 90 knot	—	12 pole permanent magnet rotor a.c. generator	—	Indicator calibrated in knot Indicator calibrated in mph Indicator calibrated in km/h
	W1400 W1402 W1404	Three cup rotor and direction vane	5–90 knot 6–100 mph 9–160 km/h	—	—	From ± ½ knot at 5 knot to ± 2 knot at 90 knot	—	12 pole permanent magnet rotor a.c. generator	Magslip	Velocity indicator calibrated in knot Velocity indicator calibrated in mph Velocity indicator calibrated in km/h
	W1500	Dines pressure tube and direction vane	0–90 knot 0–110 mph 0–180 km/h	—	—	± 1 knot at velocities above 3 knot	—	Pressure tube	Mechanical coupling	Chart recorder on alternative knot, mph or km/h paper.
	T16110/1	Three cup rotor	0·2–15 m/s 0·6–50 ft/s	—	—	—	—	Photo-electric	—	T16130 amplifier and five figure counter.
	W1710	Three cup rotor	0–80 mph 0–120 km/h 0–35 m/s 0–12 Beaufort	—	—	± 2 mph ± 3 km/h	—	Permanent magnet-drag cup	—	Hand-held instrument–integrated dial
Climatronics Corp. 1324 Motor Parkway, Hauppage, New York 11787 USA Telephone: (516) 234 2772	WS-10	Three cup rotor (vinyl or stainless steel cups)	0.5–125 mph	—	2.5 ft of flow maximum for 63% recovery (vinyl). 8 ft maximum (stainless)	The greater of ±1% or ±0.15 mph	−45.5 to 51.67	Photo-chopper system	—	Via D/A converter (translator model WT-10 to roll chart recorders; Model WR-10 for wind speed only *or* Model WR-11 for direction only *or* Model WR-12 for speed and direction
	WD-10	Direction vane	0.25 mph threshold	—	2.5 ft of flow maximum for 63% recovery	±2.5°	−45.5 to +51.67	—	Electrostatic	
	Wind Mark III	Three cup rotor and direction vane	0–50 mph *or* 0–100 mph; 0.75 mph threshold	—	8 ft of flow maximum	Speed, ±5 mph *or* 2% (system); direction ±2% (system)	—	Photo-chopper system	Precision potentiometer	Dual channel roll chart recorder

Manufacturer	Model	Type	Wind speed range	Maximum wind speed	Response	Accuracy	Temperature range (°C)	Speed transducer	Direction transducer	Output instrument
	F460	Three cup rotor and direction vane	0.5 knot threshold	—	5 ft of flow maximum (speed), 3.7 ft max (direction)	Speed, ±0.13 knot *or* 1%, direction ±3°	—	—	—	Modular rack mounted signal processor with individual translators for each of several speed or direction channels
Climet Instruments Co, PO Box 151, 1320 W. Colton Ave, Redlands, California 92373 USA Telephone: (714) 793 2788	011-3	Three cup rotor (Lexan, Teflon or stainless steel cup assembly)	0.6 to 90 mph	110 mph	5 ft of flow for 63% recovery	±1% *or* 0.15 mph, whichever is greater	−45 to +68	Photo-chopper system	—	Signal conditioning module type 060, converting sensor signals into standardized voltage and current output. Strip chart output available when sensor output is used with electronic signal conditioning
Danjay Designs, 11 Lansdowne Road, Dry Sandford, Nr Abingdon, Oxon OX13 6EA, UK Telephone: Oxford 735056	AB774S	Three cup rotor	1–120 mph; 1.5–195 km/h; 0.6–70 m/s; 0.8–105 knot	—	—	—	−30 to +70	Hall effect devices	—	AB7761 solid state with three digit liquid crystal numeric display (AB774S and AB775D)
	AB775D	Direction vane	—	—	—	—	−30 to +70	—	Conductive plastic element	
	AB771S	Three cup rotor	1–110 mph	—	—	—	−30 to +70	Solid state magnetic field sensor	—	AB7731 solid state with three digit numeric display
Deuta-Werke GmbH, Postfach 20 02 60, Paffrather Strasse 140, 5060 Bergisch Gladbach 2, West Germany Telephone: (0 22 02) 5 10 91 Telex: 887 725	Anemo	Three cup rotor	0–120 km/h; 0–80 mph; 0–35 m/s; 0–12 Beaufort	—	—	±3% at 25°C	—	Permanent magnet drag cup	—	Hand held instrument, integral dial

Fig. 7.15 Brookes and Gatehouse Model 152 three-cup anemometer.

Fig. 7.16 Casella Model W1208/2 inclined scale three-cup anemometer.

Fig. 7.17 Casella Model T 16110/1 sensitive three-cup anemometer

Manufacturer	Model	Type	Wind speed range	Maximum wind speed	Response	Accuracy	Temperature range (°C)	Speed transducer	Direction transducer	Output instrument
Dwyer Instruments Inc., Box 373, Michigan City, Indiana 46360, USA	Wind Meter	Hand held pressure tube	2–10 mph 4–66 mph	—	—	Within ½ mph on low scale, 3 mph on high scale	—	Pressure supporting ball in tube	—	Indicating; ball moving against integral scale; hand held instrument
	Wind Speed Indicator	Pressure tube and direction vane	0–80 mph, 0–12 Beaufort	—	—	—	—	Pressure tube and liquid coupling	Vane retains pressure tube in wind direction	Indicating; liquid moving along scale mounted remote from sensor
J-Tec Associates Inc, 317 7th Avenue, S.E., Cedar Rapids, Iowa 52401, USA Telephone: (319) 366 7511	VA-300	Vortex anemometer	2–135 mph	—	6 mm speed constant	±2% full scale	−30 to +71	Ultrasonic beam	—	VR-100 meter display for speed
	VA-310	Vortex anemometer	2–135 mph	—	6 mm speed constant, 10 m direction constant	Speed, ±2% full scale. Direction, ±4° at 4.5 mph, ±2° above 9 mph	−30 to +71	Ultrasonic beam	Potentiometer for analogue output	VR-110 meter display for speed and direction *or* VR-120 digital display for speed and direction
	VA-311	Vortex anemometer	2–135 mph	—	6 mm speed constant, 10 m direction constant	Speed, ±2% full scale. Direction ±4° at 4.5 mph, ±2° above 9 mph	−30 to +71	Ultrasonic beam	60 Hz synchro	
	VA-320	Vortex anemometer	2–135 mph	—	6 mm speed constant, 10 m direction constant	Speed, ±2% full scale. Direction ±4° at 4.5 mph, ±2° above 9 mph	−30 to +71	Ultrasonic beam	Potentiometer or synchro	VR-121 meter display for speed and direction *or* VR-131 digital display for wind speed and direction
Kahl Scientific Instrument Corporation, PO Box 1166, El Cajon (San Diego), California 92022, USA Telephone: (714) 444 2158 (714) 444 5944	02 AM 300 02 AM 310	Three cup rotor and direction vane	—	—	—	Better than ±5° Better than 0.5 m/s	−25 to +60	Mechanical drive (cumulative, for wind run)	Mechanical drive	Battery rewound roll-chart recorder (60 day) Hand wound roll-chart recorder (8 day)
	03 AM 010 03 AM 030 03 AM 045 03 AM 050	Three cup rotor and direction vane	0–100 mph, threshold 1.5 mph	—	—	—	—	a.c. permanent magnet generator	Potentiometer *or* dry reed switches and lamps	Dial indicators (11 cm diameter) Dial indicators (15 cm diameter) Integrated direction/speed dial Dial indicators (rectangular)
	03 AM 015 03 AM 035 03 AM 055	Three cup rotor	0–100 mph, threshold 1.5 mph	—	—	—	—	a.c. permanent magnet generator	—	Dial indicator (11 cm diameter) Dial indicator (15 cm diameter) Dial indicator (rectangular)
	03 AM 020 03 AM 040 03 AM 060	Direction vane	—	—	—	—	—	—	Potentiometer *or* dry reed switches and lamps	Dial indicator (11 cm diameter) Dial indicator (15 cm diameter) Dial indicator (rectangular)

Manufacturer	Model	Type	Wind speed range	Maximum wind speed	Response	Accuracy	Temperature range (°C)	Speed transducer	Direction transducer	Output instrument
	03 BM 010	Pressure-plate wind speed indicator	5–70 mph	—	—	—	—	Hinged pressure plate	—	Speed indicator
	03 BM 012	Pressure-plate wind speed indicator	2–25 m/s	—	—	—	—	Hinged pressure plate	—	Speed indicator
	03 BM 015	Pressure-plate wind speed indicator	5–70 mph	—	—	—	—	Hinged pressure plate	—	Speed and direction (compass)
	03AM 120	Three cup rotor and direction vane	0–70 mph and 0–100 ft/s *or* 0–60 knot and 0–30 m/s	—	1.6 ft/s threshold	±1.6 ft/s	−25 to +60	Drag-cup	Pointer indication	Hand held instrument, integral dials
Makino Ohyosokki Kenkyusho Inc., 3-19-4 Numabukuro, Nakano-Ku, Tokyo, Japan Telephone: 03-387-6241	VFO16	Direction vane	—	—	—	±3°	—	—	Photo-cell	Digital chart *or* tape recorder
	VRO 36	Direction vane	—	—	—	±3°	—	—	Potentiometer	Analogue chart *or* tape recorder
	AF750	Three cup rotor	—	—	2.5 m run of wind	±3%	—	Photo-cell	—	Digital chart *or* tape recorder *or* digital electronic counter (LC101)
	AG750	Three cup rotor	—	—	2.5 m run of wind	±3%	—	Generator	—	Analogue chart *or* tape recorder
	AC750	Three cup rotor	—	—	2.5 m run of wind	±3%	—	'Magnet contact'	—	Digital magnetic counter (KC101)
	AGP230	Three-axis propeller	—	—	2.5 m run of wind	±3%	—	Generator	—	Analogue chart *or* tape recorder
Carl Th. Malling Fabrikation ApS, Stamholmen 173, DK-2650 Hvidovre, Denmark Telephone: (01) 49 03 00 Telex: 19334	878	Five cup rotor and twin direction vane	2–40 m/s	70 m/s	0.2 s at 10 m/s	2.5%	−40 to +70 for sensor	a.c. generator (rectified)	Geared synchro rotor	Surface mounting direction and speed indicators (874) *or* panel mounting direction (87.02) and speed (873.02) indicators.
	878.2	Three cup rotor and twin direction vane	0–40 m/s	80 m/s	0.2 s at 10 m/s	2.5%	−40 to +70 for sensor	photo-electric	Geared synchro rotor	Surface mounting direction and speed indicators (874.2) *or* panel mounting direction (872.02) and speed (873.22) indicators.

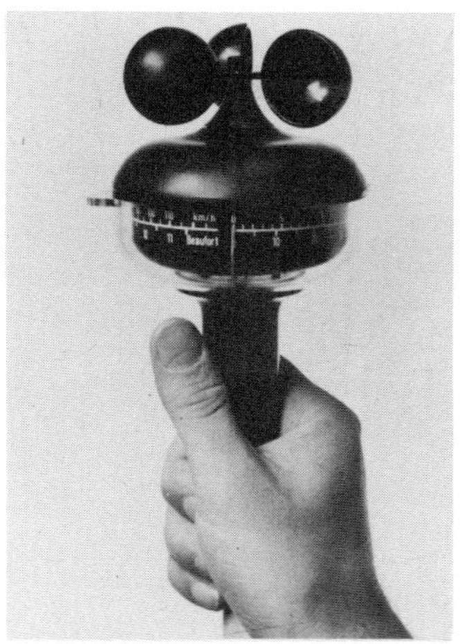

Fig. 7.18 Deuta-Werke 'Anemo' hand-held anemometer.

Manufacturer	Model	Type	Wind speed range	Maximum wind speed	Response	Accuracy	Temperature range (°C)	Speed transducer	Direction transducer	Output instrument
	881.01	Three cup rotor	0–40 m/s	80 m/s	0.2s at 10 m/s	2.5%	−40 to +70 for sensor	photo-electric	—	Surface mounting (875.12) *or* panel mounting (873.22) speed indicators.
	881.21	Direction vane	—	—	—	—	−40 to +70 for sensor	—	Geared synchro rotor	Panel mounting direction indicator (872.02)
Maximum Inc., 42 South Ave, Natick, Massachusetts 01760, USA Telephone: (617) 785-0113	Type 41 generator	Three cup rotor	3–100 mph	—	0.5 s	±3%	−40 to +65.5	a.c. permanent magnet generator	—	VIGILANT indicator (for wind speed, with gust register) *or* RADIANT II indicator (for direction only) *or* MAESTRO II indicator (for direction and wind speed, with gust register)
		Direction vane	—	—	—	—	−40 to +65.5	—	Dry reed switches for LEDs	
Meteorology Research Inc., Box 637, Altadena, California 91001, USA Telephone: (213) 791 1901 Telex: 675421	1022	Three cup rotor and direction vane	0.5–120 mph	120 mph	less than 5 ft for 63% recovery; 3.7 ft distance constant for vane	±0.15 mph or 1%, whichever greater. ±2.5°	−40 to +50	Photo-electric	Two section potentiometer	Range of indicators, analogue recording and digital data recording.
	1074	In-line three cup rotor and direction vane	0–20/80/100 mph 0–100 mph —	—	18 ft for 63% recovery (speed), 4 ft for 50% recovery (distance) threshold 0.75 mph	±0.4 mph or 1% ±0.5 mph or 1% ±0.4 mph or 1%	−40 to +50	Photo-electric Generator Switch closure	—	
	1053	Paddle vane, with two four-blade rotors	0–20/80/100 mph	—	3 ft for 50% recovery (speed and direction) threshold 0.75 mph	±0.2 mph or 1%, whichever is greater	−40 to +50	—	—	
R. W. Munro Ltd., Cline Road, Bounds Green, London N11 2LY, UK Telephone: 01 368 4422 Telex: 24130	IM144	Direction vane	—	—	—	—	—	—	Magslip transmitter	IM174 Roll chart recorder and IM120A indicators AB7761 liquid crystal display. AB7731 LED display for 16 cardinal points.
	AB775D	Direction vane	—	—	—	—	−30 to +70	—	Conductive plastic element	
	AB772D	Direction vane	—	—	—	—	−30 to +70	—	Solid state magnetic field sensor	
	IM119 IM119A	Three cup rotor	3–90 knot	—	—	±1 mph, 10–40 mph ±2.5 mph, 40–60 mph ±3 mph, 60–100 mph	—	Geared odometer	—	Vertical window, nautical miles or kilometres or miles run of wind counter. Inclined window, nautical miles or kilometres or miles run of wind counter

Manufacturer	Model	Type	Wind speed range	Maximum wind speed	Response	Accuracy	Temperature range (°C)	Speed transducer	Direction transducer	Output instrument
	IM124	Three cup rotor	0–100 mph or equivalent knot, km/h or m/s	—	—	—	—	12 pole permanent magnet rotor a.c. generator	—	IM124 moving coil d.c. voltmeter (mph, knot, km/h or m/s) (up to 6) *and/or* IM173 velocity roll chart recorder *and/or* 1A-613 series preset wind alarm
	IM146	Three cup rotor and direction vane (in-line)	0–100 mph or equivalent knot, km/h or m/s	—	—	—	—	12 pole permanent magnet rotor a.c. generator	Magslip transmitter	IM175 combined roll chart recorder for velocity and direction (mains supply via step-down transformer) *and/or* up to 6 IM120A and IM124 indicators
	IM183A	Three cup rotor and direction vane (in-line)	0–100 mph or equivalent knot, km/h or m/s	—	—	—	—	12 pole permanent magnet rotor a.c. generator	Mechanical linkage	IM184 combined roll chart recorder for velocity and direction (no mains supply required)
	IM206	Three cup rotor (IM204) and direction vane (IM 205) (in-line)	0–100 mph or equivalent knot, km/h or m/s	—	'Lightweight'	—	—	Generator with nominal 30 volt output at 1000 rev/min *or* cam-contact every 49 revolutions	Magslip transmitter	For use with either indicating or recording instruments
	IM208	Three cup rotor (IM204) and damped direction vanes (IM207) (in-line)	0–100 mph or equivalent knot, km/h or m/s	—	Damped direction vanes give slow response and retain sensitivity in light winds	—	—	Generator with nominal 30 volt output at 1000 rev/min *or* cam-contact every 49 revolutions	Contacts (MkC) *or* magslip (MkE)	For use with either indicating or recording instruments

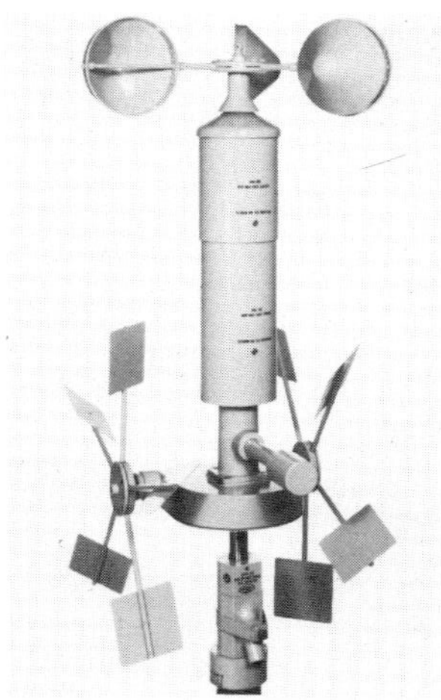

Fig. 7.19 Munro Model Im 208 three-cup anemometer and damped vane.

Fig. 7.20 Munro Model IM 159 hand-held anemometer.

Manufacturer	Model	Type	Wind speed range	Maximum wind speed	Response	Accuracy	Temperature range (°C)	Speed transducer	Direction transducer	Output instrument
	Dines Mark II	Pressure tube and direction vane	0–110 mph or equivalent knot, km/h or m/s *or* up to 200 mph	200 mph	—	—	—	Pressure tube (Met. Ref. 24)	Direct drive (Met. Ref. 25)	Flat drum chart recorder; conversion for distant reading is available (IM181)
	Dines 'Distant Reading'	Pressure tube and direction vane	0–110 mph or equivalent knot, km/h or m/s, *or* up to 200 mph	200 mph	—	—	—	Magslip transmitter (Met. Ref. 592b)	Selsyn transmitter (Met. Ref. 592a)	Flat drum chart recorder (Met Ref. 592c) with Magslip and Selsyn receivers
	IM159	Three cup rotor	0–60 mph 0–60 knot 0–30 m/s Beaufort	—	—	—	—	Permanent magnet drag cup	—	Hand held instrument, integral dial
E. Schiltknecht Ing. SIA, CH 8625 Gossau ZH, Switzerland Telephone: 01 935 2121 Telex: 75 759	655	Three cup rotor	0.6–50 m/s	50 m/s	approx. 0.5s	2%	−40 to +60	a.c. generator	—	Indicating instrument (type 811) *or* storm warning unit with max-min contact *or* flow analyser (type 655–825 m) for wind frequency analysis (6 ranges available)
	655–660 windmaster mark II	Three cup rotor and direction vane	0.6–50 m/s	50 m/s	approx. 0.5s	2%	−40 to +60	a.c. generator	—	Indicating instruments *and/or* roll chart recorder (various options available, recorded as additional digits to 655–660 model code)
Sign X Laboratories Inc., Essex, Connecticut 06426, USA Telephone: (203) 767 1700	702C	—	—	—	—	—	—	—	—	Wind spectrum analyser with 10 storage registers for individual wind speed bands
	702C-AO	—	—	—	—	—	—	—	—	As 702C, with mean wind run register added
	703A	—	—	—	—	—	—	—	—	As 702C with 20 storage registers
	703A-AO	—	—	—	—	—	—	—	—	As 703A, with mean wind run register added
	703B	—	—	—	—	—	—	—	—	As 703A, with option of using 10 registers for each of two anemometers
	801B	—	—	—	—	—	—	—	—	Aeolian odometer, recording miles of wind run
	901A	—	—	—	—	—	—	—	—	Peak Gust recorder, giving counter, binary or BCD output
	1000A	—	—	—	—	—	—	—	—	Aeolian Adometer and Peak Gust recorder (models 801B and 901A) combined in one case.

Manufacturer	Model	Type	Wind speed range	Maximum wind speed	Response	Accuracy	Temperature range (°C)	Speed transducer	Direction transducer	Output instrument
R. A. Simerl Instrument Division, 238 West Street, Annapolis, Maryland 21401, USA Telephone: (301) 849 8667	BT	Vane rotor	0–35/70 mph	—	5 mph threshold	—	—	—	—	Hand held instrument, integral dial
	BTC	3 cup rotor	0–30/60 knot							
	BX	Vane rotor	0–15/30 knot	—	5 mph threshold	—	—	—	—	Hand held instrument, integral dial
	BXC	3 cup rotor								
	M/S	Vane rotor	0–15/30 knot	—	5 mph threshold	—	—	—	—	Hand held instrument integral dial
	M/SC	3 cup rotor	0–7.6/15 m/s							
	K/K	Vane rotor	0–30/60 knot	—	5 mph theshold	—	—	—	—	Hand held instrument integral dial
	K/KC	3 cup rotor	0–56.5/113 km/h							
	SS	Vane rotor	0–6/15 m/s	—	5 mph threshold	—	—	—	—	Hand held instrument integral dial
	SSC	3 cup rotor	0–15/35 mph							
	KBM	Vane rotor	0–15/30 knot	—	5 mph threshold	—	—	—	—	Hand held instrument integral dial
	KBMC	3 cup rotor	0–7.7/15 m/s							
	RK-7	Vane rotor	0–35/100 mph	—	5 mph threshold	—	—	—	—	Remote indicating instruments R-77 and R-77C have larger meters RA and RAC have an alarm trip setting
	RK-7C	3 cup rotor	0–30/80 knot							
	RK-7KK	Vane rotor	0–60/120 km/h	—	5 mph threshold	—	—	—	—	
	RK-7KKC	3 cup rotor	0–30/60 knot							
	RS	Vane rotor	0–35/70 knot	—	5 mph threshold	—	—	—	—	
	RSC	3 cup rotor								
	R-77	Vane rotor	0–35/100 mph	—	5 mph threshold	—	—	—	—	
	R-77C	3 cup rotor	0–30/80 knot							
	RA2	Vane rotor	0–60 mph	—	5 mph threshold	—	—	—	—	
	RA2C	3 cup rotor								

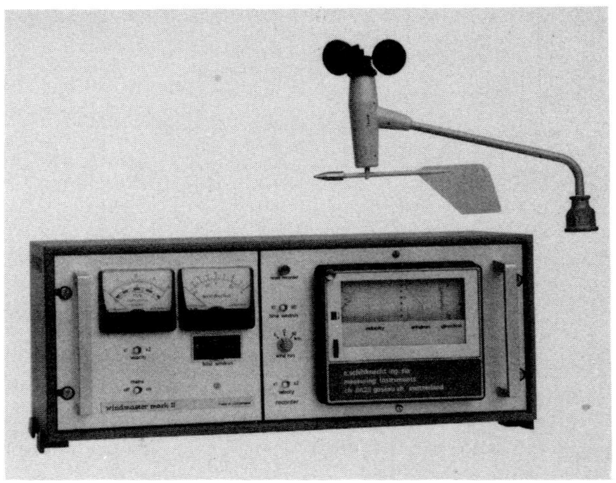

Fig. 7.21 Schiltknecht Model 655–660 l-ner/2 m windmaster mark II

Fig. 7.22 Simerl Model BTC three-cup hand-held anemometer.

Fig. 7.23 Simerl Model K/K hand-held vane anemometer.

Manufacturer	Model	Type	Wind speed range	Maximum wind speed	Response	Accuracy	Temperature range (°C)	Speed transducer	Direction transducer	Output instrument
	WS-77	Three cup rotor	0–100 mph	—	—	—	—	a.c. generator	—	Roll chart recorder
	WS-78	Three cup rotor and direction vane	0–100 mph	—	—	—	—	a.c. generator	Potentiometer	Twin roll chart recorder
Société Précis Mécanique, 14 Rue Denis-Papin, 95870 Bezons, France Telephone: 982 25-55	GPM 22 / GPM 23	Direction vane	—	—	10° maximum dead angle	—	—	—	Wire wound potentiometer / Plastic film potentiometer	Indicating instrument, roll chart recorder, or magnetic tape recorder
	APM 11		1–50 m/s					d.c. generator voltage proportional to speed (1.29 volt at 50 m/s)		
	APM 22	Three cup rotor	0.8–50 m/s	—	'Immediate'	1.5%	up to 70	d.c. generator voltage proportional to speed (14.3 volt at 50 m/s)	—	Indicating instrument, roll chart recorder or magnetic tape recorder
	APM 33		0.3–50 m/s					Photon coupled interrupter		
	AGPM 12		1–50 m/s		As for GPM 22 and APM 11			As for APM 11	As for GPM 22	
	AGPM 22	Three cup rotor and direction vane	0.8–50 m/s	—	As for GPM 22 and APM 22	1.5% on speed	up to 70 for three cup rotor	As for APM 22	As for GPM 22	Indicating instrument, roll chart recorder or magnetic tape recorder
	AGPM 23		0.3–50 m/s		As for GPM 23 and APM 33			As for APM 33	As for GPM 23	
Taylor Instrument, Consumer Products Division, Sybron Corporation, Arden, North Carolina 28704, USA	3105 / 3105A	Three cup rotor and direction vane	0–25/100 mph m/s and Beaufort range	—	—	Correction chart supplied	Up to 66	—	—	Local combined indication of speed and direction (single dial)
	3107	Three cup rotor	0–25/100 mph	—	—	Correction chart supplied	Up to 66	—	—	Local indication of speed
	3120	Three cup rotor and direction vane	0–100 mph	—	—	Correction chart supplied	Up to 66	—	—	Remote indication of speed and direction (two instruments in a single panel)
	3121	Three cup rotor	0–100 mph	—	—	Correction chart supplied	Up to 66	—	—	Remote indication of speed

Manufacturer	Model	Type	Wind speed range	Maximum wind speed	Response	Accuracy	Temperature range (°C)	Speed transducer	Direction transducer	Output instrument
Vaisala OY, PL 26, SE-00421, Helsinki 42, Finland Telephone: 358 0893 933 Telex: 122832	WAA 12	Three cup rotor	Threshold 0.4 m/s range 0–75 m/s	—	1.5 m for 63% recovery	±1% or 0.1 m/s below 20 m/s ±2% or 0.4 m/s from 20 to 75 m/s	−40 to +55	LED/Photo-transistor (chopper)	—	Analogue display of speed and direction (WAD 11), slave unit to WAD 11 (WAD 12), LED digital display (WAD 13), roll chart recorder (WAR 12) or digital averaging unit (WAI 12). Also as part of MILOS auto. weather station
	WAV 12	Direction vane	Threshold 0.3 m/s	—	1 m for 63% recovery	5.63° resolution	−40 to +55		LED/Photo transistor (code disc)	
Vector Instruments, 113 Marsh Road, Rhyl, Clwyd LL18 2AB, UK Telephone: (0745) 50700	W200	Wind vane	0.5 knot threshold	over 100 knot	less than 1 s at 5 m/s	±2° following averaging	−25 to +55	—	Reed switch potentiometer with d.c. analogue output	Wind speed and direction indicators in carrying case, type D500 (AW1500) *or* Anemograph set including roll chart recorder, type R500 (AWR500) *or* mobile laboratory set with outputs for recorder, type D600S (AW1600S1)
	A100	Three cup rotor	0.5–100 knot	over 100 knot	1 s at 5 m/s (fast response A100E type is available)	2% ±0.1 m/s after calibration	−60 to +55 (in non-icing conditions)	Photo-electric, linear d.c. analogue output		
	A100K, A100M, A100S	Three cup rotor	0.5–100 knot	over 100 knot	1 s at 5 m/s	2% ±0.1 m/s after calibration	−60 to +55	Photo-electric with a.c. digital output (10 pulses/knot, 10 pulses/m/s, 1 pulse/1.25 m)		Indicating and recording units, speed and direction alarm annunciator with memory, long period averaging unit
	A100R	Three cup rotor	0.3–50 m/s	over 50 m/s	1 s at 5 m/s	2% ±0.1 m/s after calibration	−30 to +55	Mercury wetted reed switch	—	6 digit electromagnetic counter, type EMC1
	W200G	Wind vane	0.5 knot threshold	over 100 knot	less than 1 s at 5 m/s	±2% following averaging	−25 to +55	—	Reed switch potentiometer with digital output	—

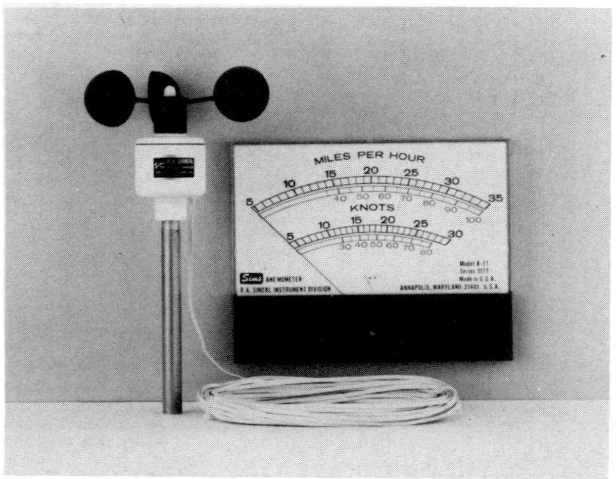

Fig. 7.24 Simerl Model R–77C three-cup anemometer and indicating meter.

Fig. 7.25 Vector Instruments three-cup anemometer and wind vane.

Manufacturer	Model	Type	Wind speed range	Maximum wind speed	Response	Accuracy	Temperature range (°C)	Speed transducer	Direction transducer	Output instrument
	SRW1	wind vane	Threshold 1.5 m/s (with marine seal)	over 50 m/s	±3° lag	±5° using time averaging	−25 to +55	—	Reed switches, self referencing to earth's magnetic field	d.c. analogue output
	SRW1G		0.75 m/s (without seal)	over 25 m/s	±2° lag					a.c. digital output
Thomas Walker & Son Ltd, 58 Oxford St., Birmingham B5 5NX, UK Telephone: 021 643 5474 Telex: 337124	MH1	Three cup rotor and direction vane	Threshold 1.5 m/s, 0–50/80 knot, 0–30 m/s, 0–80 mph	—	—	Approx ±2% full scale deflection	−20 to +50	Light emitting diode and photochopper	Desynn system	Indicating speed and direction instruments
	WG1	Three cup rotor	Threshold 3 m/s, 0–50 knot, 0–30 m/s and 0–80 mph	—	—	Approx ±2% full scale deflection	−20 to +50	a.c. generator	—	Indicating instruments in knot/Beaufort scale (WG2), mph (WG3) or m/s/Beaufort scale (WG4)
	XWD1	Direction vane	—	—	—	—	−20 to +50	—	Desynn system	Indicating instruments in cardinal points (WD2) or 0-180-0 (WD4), or alarm control box
Weather Measure Corporation, PO Box 41257, Sacramento, California 95841, USA Telephone: (916) 481 7565 Telex: 377-310	W103-3L	Three cup rotor, metallized butyrate	0.6–100 mph		5 ft distance constant			Frictionless HF tachometer incorporating solid state oscillator		
	W103-3SS	Three cup rotor stainless steel	0.9–100 mph	—	14.3 ft distance constant	±1% or 0.15 mph, whichever is greater	−34 to +60		—	
	W103B-3L	Three cup rotor metallized butyrate	0.6–100 mph		5 ft distance constant			d.c. generator		A range of translators for HF tachometer or d.c. generator. Roll chart recorders, mains or battery powered. Also in W161 wind alarm/control system (W103B-3SS)
	W103B-3SS	Three cup rotor stainless steel	0.9–100 mph		14.3 ft distance constant					
	W104	Lightweight direction vane	0.75 mph threshold	—	3.5 ft distance constant	0.72° resolution	Optional heater for cold climates	—	Low torque potentiometer with dual wipers.	
	W104-SC								Sin–cos potentiometer	

Manufacturer	Model	Type	Wind speed range	Maximum wind speed	Response	Accuracy	Temperature range (°C)	Speed transducer	Direction transducer	Output instrument
	W102-P/AC W102-P/DC W102-P/HF	Four bladed propeller and integral vane	1–200 mph	200 mph	6.2 ft distance constant at 30 mph	±1% below 25 mph ±3% above 25 mph	—	a.c. generator d.c. generator HF tachometer	Dual-wiper potentiometer	A range of translators, analogue and digital indicators, and roll chart recorder
	W102-SY/AC W102-SY/DC							a.c. generator d.c. generator	Selsyn motor	
	W200S	Three cup rotor	0–100 mph	—	—	—	—	2 pole a.c. generator	—	Indicating instruments or roll chart recorder
	W200D	Direction vane	—	—	—	—	—	—	Potentiometer	
	W200-SD	Three cup rotor and direction vane	0–100 mph	—	—	—	—	2 pole a.c. generator	Potentiometer	
	W163-1/6E W163-1E W163-100M	Three cup rotor	2.5 mph threshold	—	—	—	—	Contact each 1/6 mile Contact each mile Contact each 100 m	—	Run of wind or mean wind speed indicator
	W164-B/E W164-B/M	Three cup rotor (heavy duty)	2.5 mph threshold	—	—	—	—	Contact each 1/6 and 1 mile Contact each 100 m	—	Integral 5 digit mile register Integral 5 digit 100 m register
	W300	Three cup rotor	—	—	Equivalent to current start of art wind generator	±6.5% of reading at 95% confidence level (system)	—	d.c. generator which responds to wind in similar way to selected wind generator	—	Removable memory cell containing kW h prediction for wind generator
Westberg MFG Co., 3400 Westach Way, Sonoma, California 95476 USA Telephone: (707) 938 2121 (707) 938 4723	R2	Four cup rotor	0–50/80/100 mph	—	'Immediate'	±2%	−17.8 to +60	Magnetic switch with sealed contacts	—	Indicating instruments and/or roll chart recorder; 110 volt a.c. supply required
	V2	Direction vane	—	—	'Immediate'	±2%	−17.8 to +60	—	—	Also for warning systems
	R7G	Four cup rotor	5–50 mph *or* 5–80 mph	—	—	—	—	—	—	For wind alarm systems, with 12 volt d.c. or 115 volt a.c. supply
Wind Power Systems Inc., PO Box 17323, San Diego, California 92117, USA Telephone: (714) 452 7040	WPS Windometer	Four cup rotor	—	—	—	—	−34 to +60	—	—	CMOS memory register and solid state LED display giving run of wind

Manufacturer	Model	Type	Wind speed range	Maximum wind speed	Response	Accuracy	Temperature range (°C)	Speed transducer	Direction transducer	Output instrument
Wittich & Visser, Postbus 17339, The Hague, Netherlands Telephone: 070-645237	PRR1 PRR2 PRR4	Direction vane	1.5 m/s normal threshold, but can be made higher by use of special brake	—	—	—	—	—	Single reed relay (2×180° sectors) Two reed relays (2×120° sectors) Four reed relays (4×90° sectors)	Suitable for connection to wind speed and direction alarm systems Model 8WA *or* Model 8WA1-A (single stage alarm) *or* Model 8WA2-A (two stage alarm)
	PAR PAF5-50 PAF12-50	Three cup rotor	—	—	—	—	−30 to +40	Permanent magnet generator Photocell, 5 volt, 50 Hz at 30 m/s Photocell, 12 volt, 50 Hz at 30 m/s	—	
R. M. Young Co., 2801 Aero-Park Drive, Traverse City, Michigan 49684, USA	12302	Direction vane	Threshold less than 0.4 m/s	—	1 m for 50% recovery	1%	−50 to +50	—	Conductive plastic potentiometer (Model 12302C gives 540° azimuth range instead of normal 360°)	A range of translators and recorders are available
	12102 12102D	Three cup rotor	Threshold 0.35–0.45 m/s Threshold 0.2–0.3 m/s	50 m/s	2.4 m for 63% recovery	1%	−50 to +50	Tachometer generator Photo-transistor light-chopper	—	
	12002	Three cup rotor and direction vane	Threshold approx. 0.4 m/s range 0–50 m/s	50 m/s	As for 12102 and 12302	1%	−50 to +50	As for 12102/12102D	As for 12302/12302C	
	21003	Propeller bivane (four blade)	Threshold of both propeller and vane, 0.1–0.2 m/s range 0–30 m/s	—	Vane, 1 m for 50% recovery; propeller, 1 m for 63% recovery	—	—	d.c. tachometer generator	Conductive plastic potentiometers (Model 21003C gives 540° azimuth range instead of normal 360°)	Power supply translator and 3 m indicator panel available

Manufacturer	Model	Type	Wind speed range	Maximum wind speed	Response	Accuracy	Temperature range (°C)	Speed transducer	Direction transducer	Output instrument
	35003	Propeller vane (four blade)	Threshold 0.1–0.2 m/s for propeller and vane, 0–30 m/s range	—	1 m for 63% recovery of propeller, 1.2 m for 50% recovery of vane	1%	−50 to +50	d.c. generator, voltage proportional to speed	Geared conductive plastic potentiometer (Model 35003C gives 540° azimuth range instead of normal 360°)	Translators and recorders are available
	17003	Direction bivane	Threshold 0.1–0.2 m/s	—	1 m for 50% recovery	—	—	—	Conductive plastic potentiometers (Model 17003C gives 540° azimuth range instead of normal 360°)	Power supply translator available
	27103	Propeller (four blade)	Threshold 0.1–0.2 m/s for 23 cm × 30 cm propeller	—	1 m for 63% recovery	—	—	d.c. tachometer generator or photo-transistor light chopper	—	—
			0.2–0.3 m/s for 19 cm × 30 cm propeller	—	0.8 m for 63% recovery					

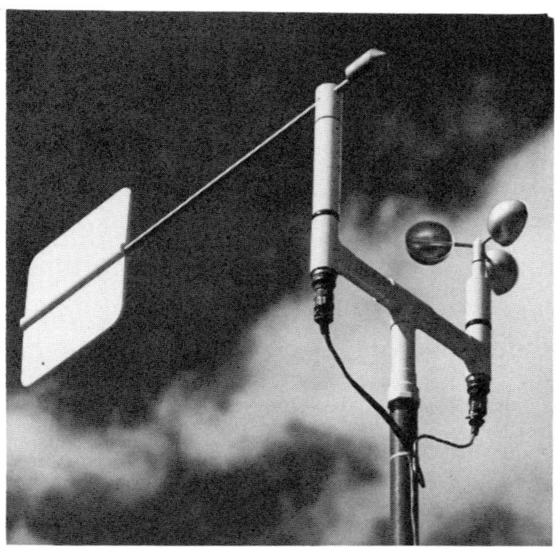

Fig. 7.26 R.M. Young Model 12001 Gill microvane and three-cup anemometer.

Fig. 7.27 R.M. Young Model 35003 Gill propeller vane.

Manufacturer	Model	Type	Wind speed range	Maximum wind speed	Response	Accuracy	Temperature range (°C)	Speed transducer	Direction transducer	Output instrument
	27004	Three-axis propeller (four blade)	Standard propeller, 0.1–0.2 m/s threshold, 0–22 m/s range. Smaller propeller, 0.2–0.3 m/s threshold, 0–30 m/s range	—	Standard propeller 1 m for 63% recovery, smaller propeller, 0.8 m for 63% recovery	1% to 10% approx. depending upon wind angle	−50 to +50	d.c. tachometer generator	—	Indicator–translator unit
	6301	Direction vane	Threshold less than 0.7 m/s	—	Damping ratio 0.37	—	—	—	Conductive plastic potentiometer	
	6101	Three cup rotor	Threshold less than 0.7 m/s range 0–54 m/s	—	3.7 m for 63% recovery	—	—	d.c. tachometer generator	—	
	6106	Three cup rotor	Threshold less than 0.6 m/s range 0–54 m/s	—	3.2 m for 63% recovery	—	—	Wirewound potentiometer making one revolution each 10 miles	—	Roll chart recorder
	6006	Three cup rotor and direction vane	As 6106 and 6301	—	As 6106 and 6301	—	—	As 6106	As 6301	Various roll chart recording systems
	6001	Three cup rotor and direction vane	As 6101 and 6301	—	As 6101 and 6301	—	—	As 6101	As 6301	Various roll chart recording systems
	8002	Propeller vane (three blade)	Threshold less than 0.4 m/s range 0–54 m/s	—	2.4 m for 63% recovery	—	—	d.c. tachometer generator	Geared conductive plastic potentiometer (Model 8002C gives 540° azimuth instead of normal 360°)	Various roll chart recording systems

Unit conversion table

Class	To convert	into	multiply by
Length	metres	feet	3.281
	millimetres	inches	0.0394
Volume	cubic metres (m^3)	Imp. gallons	220.5
	cubic metres (m^3)	US gallons	264.2
	cubic metres (m^3)	litres	1000
Mass	kilograms	pounds	2.205
Speed	m/s	mph	2.237
	m/s	knots	1.942
Flow rate	m^3/h	Imp. gall/min	3.68
	m^3/h	US gall/min	4.40

Index